Sensors and Transducers

A Guide for Technicians

Sensors and Transducers
A Guide for Technicians

Ian R. Sinclair

BSP PROFESSIONAL BOOKS

OXFORD LONDON EDINBURGH

BOSTON PALO ALTO MELBOURNE

First published by BSP Professional
Books 1988

British Library
Cataloguing in Publication Data
Sinclair, Ian R. (Ian Robertson), *1932–*
    Sensors and transducers.
    1. Electronic equipment. Transducers
    I. Title
    621.37'9

ISBN 0–632–02069–5

BSP Professional Books
A division of Blackwell Scientific
    Publications Ltd
Editorial Offices:
Osney Mead, Oxford OX2 0EL
    (Orders: Tel. 0865 240201)
8 John Street, London WC1N 2ES
23 Ainslie Place, Edinburgh EH3 6AJ
3 Cambridge Center, Suite 208, Cambridge,
    MA 02142, USA
667 Lytton Avenue, Palo Alto, California
    94301, USA
107 Barry Street, Carlton, Victoria 3053,
    Australia

Set by DP Photosetting, Aylesbury, Bucks
Printed and bound in Great Britain by
Mackays of Chatham PLC, Chatham, Kent

Contents

Preface

The purpose of this book is to explain and illustrate the use of sensors and transducers associated with electronic circuits. The steady spread of electronic circuits into all aspects of life, but particularly into all aspects of control technology, has greatly increased the importance of sensors which can detect, as electrical signals, changes in various physical quantities. In addition, the conversion by transducers of physical quantities into electronic signals and vice versa has become an important part of electronics.

Because of this, the range of possible sensors and transducers is by now very large, and most textbooks that are concerned with the interfaces between electronic circuits and other devices tend to deal only with a few types of sensors for specific purposes. In this book, you will find described a very large range of devices, some used industrially, some domestically, some employed in teaching to illustrate effects, some used only in research laboratories. The important point is that the reader will find reference to a very wide range of devices, much more than it would be possible to present in a more specialised text.

In addition, I have assumed that the physical principles of each sensor or transducer will not necessarily be familiar. To be useful, a book of this kind should be accessible to a wide range of users, and since the correct use of sensors and transducers often depends critically on an understanding of the physical principles involved, these principles have been explained in as much depth as is needed. I have made the reasonable assumption that electrical principles will not require to be explained in such depth as the principles of, for example, relative humidity. In order for the book to be as serviceable as possible to as many readers as possible, the use of mathematics has been avoided unless absolutely essential to the understanding of a device. I have taken here as my guide the remark by Lord Kelvin that if he needed to use mathematics to explain something it was probably because he didn't really understand it. The text should prove useful to anyone who encounters sensors and transducers, whether from the point of view of specification, design, servicing, or education.

I am most grateful to RS Components for much useful and well-organised information, and to Bernard Watson, of BSP Professional Books, for advice and encouragement.

Ian Sinclair
April 1988

Introduction

A sensor is a device that detects or measures a physical quantity, and in this book the types of sensors that we are concerned with are the types whose output is electrical. A transducer is a device which converts energy from one form into another, and here we are concerned only with the transducers in which one form of energy is electrical. The differences between sensors and transducers are often very slight. A sensor is performing a transducing action, and the transducer must necessarily sense some physical quantity. The shade of difference lies in the efficiency of energy conversion. The purpose of a sensor is to detect and measure, and whether its efficiency is 5% or 0.1% is almost immaterial, provided the figure is known. A transducer, by contrast, is intended to convert energy, and its efficiency is important, though in some cases it may not be high. Linearity of response, important for a sensor, may be of much less significance for a transducer. The basic principles that apply to one, however, must apply to the other, so that the descriptions that appear in this book will apply equally to sensors and to transducers.

The organisation of the book is in general by the physical quantity that is sensed or converted. This is not a perfect form of organisation – none is – because there are many 'one-off' devices that sense or convert for some unique purpose, and these have had to be gathered up in an 'assortment' chapter. Nevertheless, by grouping devices according to the sensed quantity, it is much easier for the reader to find the information that is needed, and that is the guiding principle for this book. In addition, some of the devices that are dealt with early in the book are those which form part of other sensing or transducing systems that appear later. This avoids having to repeat a description, or refer forward for a description.

Several points should be noted at this stage, to avoid much tedious repetition in the main body of the book. One is that a fair number of physical effects are sensed or measured, but have no requirement for transducers – we do not, for example, generate electricity from earthquake shocks. A second point is that the output from a sensor, including the output from electronic circuits connected to the sensor, needs to be proportional in some way to the effect that is being sensed, or at least to bear some simple mathematical relationship to the quantity. This means that if the output is to be used for measurements, then some form of calibration can be carried out. It also implies that the equation that connects the electrical output with the input that is being sensed contains various constants such as mass, length, resistance and so on. If any of these quantities is varied at any time, then recalibration of the equipment will be necessary.

Another point that we need to be clear about is the meaning of resolution as applied to a sensor. The resolution of a sensor measures its ability to detect a change in the sensed quantity, and is usually quoted in terms of the smallest change that can be detected. In some cases, resolution is virtually infinite, meaning that a small change in the sensed quantity will cause a small change in the electrical output, and these changes are detectable to the limits of our measuring capabilities. For other sensors, particularly when digital methods are used, there is a definite limit to the size of change that can be either detected or converted. It is important to note that very few sensing methods provide a digital output directly, and most digital outputs are obtained by converting from analogue quantities. This implies that the limits of resolution are determined by the analogue to digital conversion circuits rather than by the sensor itself. Where a choice of sensing methods exists, a method that causes a change of frequency of an oscillator is to be preferred, because frequency is a quantity that lends itself very easily to digital handling methods with no need for other analogue to digital conversion methods.

The sensing of any quantity is liable to error, and the errors can be static or dynamic. A static error is the type of error that is caused by reading problems, such as the parallax of a needle on a meter scale, which causes the apparent reading to vary according to the position of the observer's eye. Another error of this type is the interpolation error, which arises when a needle is positioned between two marks on a scale, and the user has to make a guess as to the amount signified by this position. The amount of an interpolation error is least when the scale is linear. One distinct advantage of digital readouts is that neither parallax nor interpolation errors exist. The other form of error is dynamic, and a typical error of this type is a difference between the quantity as it really is and the amount that is measured, caused by the loading of the measuring instrument itself. A familiar example of this is the false voltage reading measured across a high-resistance potential divider with a voltmeter whose input resistance is not high enough. All forms of sensors are liable to dynamic errors if they are used only for sensing, and to both dynamic and static errors if they are used for measurement.

Finally, two measurable quantities can be quoted in connection with any sensor or transducer. These are responsivity and detectivity, and though the names are not necessarily used by the manufacturer of any given device, the figures are normally quoted in one form or another. The responsivity is:

$$\frac{\text{output signal}}{\text{input signal}}$$

which will be a measure of transducing efficiency if the two signals are in comparable units (both in watts, for example), but which is normally expressed with very different units for the two signals. The detectivity is defined as:

$$\frac{\text{S/N of output signal}}{\text{size of input signal}}$$

where S/N has its usual electrical meaning of signal to noise ratio. This latter definition can be reworked as:

$$\frac{\text{responsivity}}{\text{output noise signal}}$$

if this makes it easier to measure.

Chapter One
Strain and Pressure

Mechanical strain

The words stress and strain are often confused in everyday life, and a clear definition is essential at this point. Strain is the result of stress, and is a fractional change of the dimensions of an object. By fractional, I mean that the change of dimension is divided by the original dimension, so that in terms of length, for example, the strain is the change of length divided by the original length. This is a quantity which is a pure number, one length divided by another, having no physical dimensions. Strain can be defined for area or for volume in a similar way as change divided by original quantity. A stress, by contrast, is a force divided by an area. As applied to a wire or a bar in tension, for example, the tensile stress is the applied force divided by the area over which it is applied, the area of cross section of the wire or bar. For materials which can be compressed, the bulk stress is the force per unit area, which is identical to pressure applied, and the strain is the change of volume divided by the original volume. The most common strain transducers are for tensile mechanical strain.

Sensing tensile strain involves the measurement of very small changes of length of a sample. This is complicated by the effect of changes of temperature, which produce expansion or contraction that for the changes of 0°C to 30°C that we encounter in atmospheric temperature are of the same order of size as the changes caused by large amounts of stress. Any system for sensing and measuring strain must therefore be designed in such a way that temperature effects can be compensated for if necessary.

The commonest form of strain measurement uses resistive strain gauges. A resistive strain gauge consists of a conducting material in the form of a thin wire or strip which is attached firmly to the material in which strain is to be detected. This material might be the wall of a building, a turbine blade, part of a bridge, anything in which excessive stress could signal impending trouble. The fastening of the resistive material is usually by means of epoxy resins (such as 'Araldite'), since these materials are extremely strong and are electrical insulators. The strain gauge strip will then be connected as part of a resistance bridge circuit (Fig. 1.1). The effects of temperature can be mini-mised by using another identical unstrained strain gauge in the bridge as

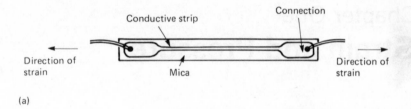

Conductive strip

Connection

Direction of
strain

Mica

Direction of
strain

(a)

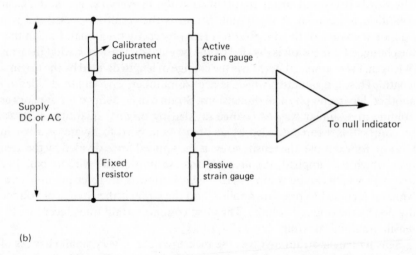

Calibrated
adjustment

Active
strain gauge

Supply
DC or AC

To null indicator

Fixed
resistor

Passive
strain gauge

(b)

Fig. 1.1 Strain gauge use. (a) Physical form of a strain gauge. (b) A bridge circuit for strain gauge use. By using an active (strained) and a passive (unstrained) gauge in one arm of the bridge, temperature effects can be compensated if both gauges are identically affected by temperature. The two gauges are usually side by side, but with only one fastened to the cause of strain.

a comparison. This is necessary not only because the material under investigation will change dimensions as a result of temperature changes, but because the resistance of the strain gauge element itself will vary. By using two identical gauges, one unstrained, in the bridge circuit, these changes can be balanced against each other, leaving only the change that is due to stress. The sensitivity of this type of gauge, often called the piezoresistive gauge, is measured in terms of the gauge factor. This is defined as the fractional change of resistance divided by the change of strain, and is typically about 2 for a metal wire gauge and about 100 for a semiconductor type.

The change of resistance of a gauge constructed using conventional wire elements (typically thin Nichrome wire) will be very small, as the gauge factor figures above indicate. Since the resistance of a wire is proportional to its length, the fractional change of resistance will be equal to the fractional change of length, so that changes of less than 0.1% need to be detected. Since the resistance of the wire element is small, of the order of an ohm or less, the actual change of resistance is likely to be very small compared to the resistance of connections in the circuit, and this can make measurements very uncertain when small strains have to be measured. The use of a semiconductor strip in place of a metal wire makes measurement much easier, because the resistance of such a strip can be considerably greater, and so the changes in resistance can be correspondingly greater. Except for applications in which the temperature of the element is high (gas-turbine blades, for example), the semiconductor type of strain gauge is preferred. Fastening is as for the metal type, and the semiconductor material is surface passivated – protected from atmospheric contamination by a layer of oxidation on the surface. This latter point can be important, because if the atmosphere around the gauge element removes the oxide layer, then the readings of the gauge will be affected by chemical factors as well as by strain, and measurements will no longer be reliable.

Piezoelectric strain gauges are useful where the strain is of short duration, or rapidly changing in value. A piezoelectric material is a crystal whose ions move in an asymmetrical way when the crystal is strained, so that an EMF is generated between two faces of the crystal (Fig. 1.2). The EMF can be very large, of the order of several kV for a heavily strained crystal, so that the gauge can be sensitive, but the output impedance is very high and capacitive (Fig. 1.3). The output is not DC, therefore, so that this type of gauge is not useful

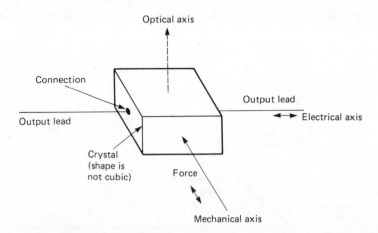

Fig. 1.2 Piezoelectric crystal principles. The crystal shape is not cubic, but the directions of the effects are most easily shown on a cube. The maximum electric effect is obtained across faces whose directions are at right angles to the faces on which the force is applied. The third axis is called the optical axis because light passing through the crystal in this direction will be most strongly affected by polarisation (see Chapter 3).

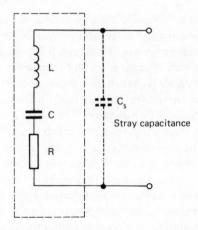

Fig. 1.3 The equivalent circuit of a crystal. This corresponds to a series resonant circuit with very high inductance, low capacitance and almost negligible resistance.

for detecting slow changes, and its main application is for acceleration sensing (see Chapter 2).

Two major problems of strain gauge elements of any type are hysteresis and creep. Hysteresis means that a graph of resistance change plotted against length change does not follow the same path of decreasing stress as for increasing stress (Fig. 1.4). Unless the gauge is over-stretched, this effect should be small, of the order of 0.025% of normal readings at the most. Over-stretching of a strain gauge will cause a large increase in hysteresis, and, if excessive, will cause the gauge to show a permanent change of length, making it useless. The other problem, creep, refers to a gradual change in the length of the gauge element which does not correspond to any change of strain in the

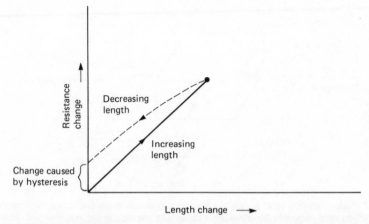

Fig. 1.4 The hysteresis effect on a strain gauge, greatly exaggerated. The graph is linear for increasing strain, but does not take the same path when the strain is decreasing. This results in the gauge having permanently changed resistance when the strain is removed.

material that is being measured. This also should be very small, of the order of 0.025% of normal readings. Both hysteresis and creep are non-linear effects which can never be eliminated but which can be reduced by careful choice of the strain gauge element material. Both hysteresis and creep increase noticeably as the operating temperature of the gauge is raised.

Interferometry

Laser interferometry is another method of strain measurement that presents considerable advantages, not least in sensitivity. Though the principles of the method are quite ancient, its practical use had to wait until suitable lasers and associated equipment had been developed, along with practicable electronic methods of reading the results. Before we can look at what is involved in a laser interferometer strain gauge, we need to understand the basis of wave interference and why it is so difficult to achieve with light.

All waves exhibit the effect that is called interference (Fig. 1.5). When two waves meet and are in phase (peaks of the same sign coinciding), then the result is a wave of greater amplitude, a reinforced wave. This is called constructive interference. If the waves are in opposite phase when they meet, then the sum of the two waves is zero, or a very small amplitude of wave, and this is destructive interference. The change from constructive to destructive

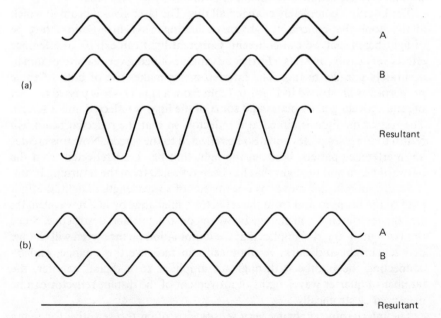

Fig. 1.5 Wave interference. When waves meet and are in phase (a), the amplitudes add so that the resultant wave has a larger amplitude. If the waves are in antiphase (b), then the resultant is zero, or of small amplitude.

interference therefore occurs for a change of phase of one wave relative to another of half a cycle. If the waves are emitted from two sources, then a movement of one source by a distance equal to half a wavelength will be enough to change the interference from constructive to destructive or vice versa. If the waves that are used have a short wavelength, then the distance of half a wavelength can be very short, making this an extremely sensitive measurement of change of distance.

The wavelength of red light is about 700 nm, that is 7^{-7} metres or 7^{-4} mm, so that a shift of half this distance between two sources of this light could be expected to cause the change between fully constructive and fully destructive interference – in practice we could detect a considerably smaller change than this maximum amount. The method would have been used much earlier if it were not for the problem of coherence. Interference is possible if the waves that are interfering are continuous over a sufficiently long period. Conventional light generators, however, do not emit waves continuously. In a light source such as a filament bulb or a fluorescent tube, each atom emits a pulse of light radiation, losing energy in the process, and then stops emitting until it has regained energy. The light is therefore the sum of all the pulses from the individual atoms, rather than a continuous wave. This makes it impossible to obtain any interference effects between two separate normal sources of light, and the only way that light interference can normally be demonstrated is by using light that has passed through a pinhole to interfere with its own reflection, with the light path difference very small.

The laser has completely changed all this. The laser gives a beam in which all the atoms that contribute light are oscillating in synchronisation; the type of light beam that we call coherent. Coherent light can exhibit interference effects very easily, and has a further advantage of being very easy to obtain in accurately parallel beams. The interferometer makes use of both of these properties as illustrated in Fig. 1.6. Light from a small laser is passed to a set of semi-reflecting glass plates and some of the light is reflected into a screen. The rest of the light is aimed at a reflector, so that the reflected beam will return to the glass plates and also be reflected to the screen. Now this creates an interference pattern between the light that has been reflected from the outward beam and the light that has been reflected from the returning beam. If the distant reflector moves by one quarter of a wavelength of light, the light path of the beam to and from the reflector will change by half a wavelength, and the interference will change between constructive and destructive. Since this is a light beam, this implies that the illumination on the screen will change between bright and dark. A photocell can measure this change, and by connecting the photocell through an amplifier to a digital counter, the number of quarter wavelengths of movement of the distant reflector can be measured electronically.

The interferometer, being very sensitive, is often too sensitive for many purposes. For example, the effect of changing temperatures is not easy to compensate for, though this can be done by using elaborate light paths in

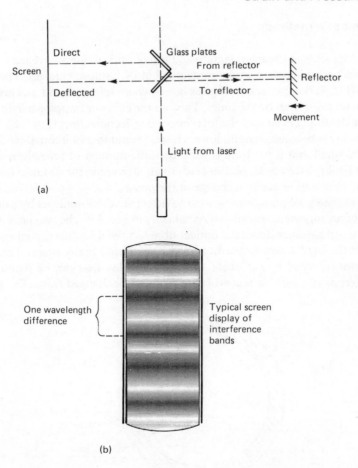

Fig 1.6 Principles of wave interferometry. The setup of laser and glass plates is shown in (a). The glass plates will pass some light and reflect some, so that both the reflector and the screen will receive some light from the laser beam. In addition, the light reflected from the reflector will also strike the screen, causing an interference pattern (b). For a movement of half of one wavelength of the reflector, the pattern will move a distance equal to the distance between bands on the screen.

which the two interfering beams have travelled equal distances, one in line with the stress and the other in a path at right angles. An advantage of this method is that no physical connection is made between the points whose distance is being measured; there is no wire or semiconductor strip joining the points, only the interferometer main body in one place and the reflector in another. The distance between the main part of the device and the reflector is not fixed, the only restraint being that the distance must not exceed the coherence distance for the laser. This is the average distance over which the light remains coherent, and is usually at least several metres for a laser source.

Fibre optic methods

Developments in the manufacture and use of optical fibres have led to these devices becoming used in the measurement of distance changes. The optic fibre (Fig. 1.7) is composed of glass layers whose refractive index is lower on the outer layer than on the inner. This has the effect of trapping a light beam inside the fibre because of the total internal reflection effect (Fig. 1.8). When a light ray is beamed straight down a fibre, the number of internal reflections will be small, but if the fibre is bent, then the number of reflections will be considerably increased, and this leads to an increase in the distance travelled by the light and hence to a change in the phase.

This change of phase can be used to detect small movements by using the type of arrangement shown diagramatically in Fig. 1.9. The two jaws will, as they move together, force the optical fibre to take up a corrugated shape in which the light beam in the fibre will be reflected many times. The extra distance travelled by the beam will cause a delay that can be detected by interferometry, using a second beam from an unchanged fibre. The sensor

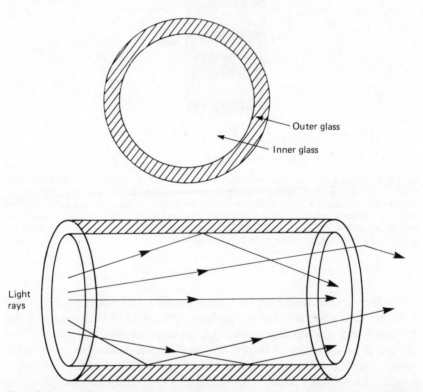

Fig. 1.7 Optical fibre construction. The optical fibre is not a single material but a coaxial arrangement of transparent glass or (less usefully) plastics. The materials are different and refract light to different extents (refractivity) so that any light ray striking the junction between the materials is reflected back and so trapped inside the fibre.

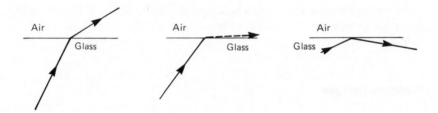

Fig. 1.8 Total internal reflection. When a ray of light passes from an optically dense (highly refractive) material into a less dense material, its path is refracted away from the original direction (a) and more in line with the surface. At some angle (b), the refracted beam will travel parallel to the surface, and at glancing angles (c), the beam is completely reflected. The use of two types of glass in an optical fibre ensures that the surface is always between the same two materials, and the outer glass is less refractive than the inner to ensure reflection.

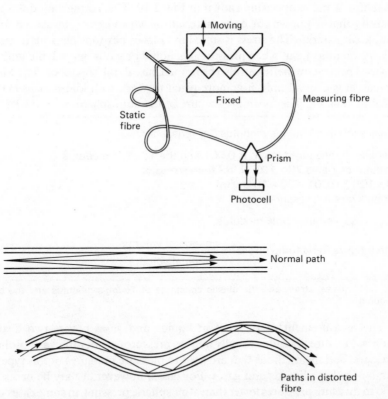

Fig. 1.9 Using optical fibres to detect small distance changes. The movement of the jaws distorts one fibre, forcing the light paths to take many more reflections and thus increasing the length of the total light path. An interference pattern can be obtained by comparing this to light from a fibre that is not distorted, and the movement of the pattern corresponds to the distortion of one fibre. The sensitivity is not so great as that of direct interferometry, and the use of fibres makes the method more generally useful, particularly in dark liquids or other surroundings where light beams could not normally penetrate.

must be calibrated over its whole range, because there is no simple relationship between the amount of movement and the amount by which the light is delayed.

Pressure gauges

Pressure in a liquid or a gas is defined as the force acting per unit area of surface. This has the same units as mechanical stress, and for a solid material, the force/area quantity is always termed stress rather than pressure. For a solid, the amount of stress would be calculated, either from knowledge of force and area of cross section, or from the amount of strain. Where the stress is exerted on a wire or girder, the direct calculation of stress may be possible, but since strain can be measured by electronic methods, it's usually easier to make use of the relationship shown in Fig. 1.10. The Young's modulus is a quantity that is known for each material, or which can be measured for a sample of material. The stress is stated in units of newtons per square metre (a large quantity), but when pressure in a liquid or gas is quoted, the units of newtons per square metre can also be termed pascals (abbreviation: Pa). Since the pascal is a small unit, it is more usual to work with kilo-pascals (kPa), equal to 1000 Pa. The 'normal' pressure of the atmosphere is 101.3 kPa.

---

stress = strain × Young's modulus (for tensile stress)

Example: If measured strain is 0.001 and the Young's modulus for the material is 20×10^{10} N/m^2 then stress is
$20 \times 10^{10} \times 0.001 = 20 \times 10^7$ N/m^2
For bulk stress (pressure), use:

   stress = strain × bulk modulus

with a figure for volume stress $= \dfrac{\text{change of volume}}{\text{original volume}}$

Fig. 1.10 Stress, strain and the elastic constants of Young's modulus and the bulk modulus.

The measurement of pressure in liquids and gases covers two distinct ranges. Pressure in liquids usually implies pressures greater than atmospheric pressure, and the methods that are used to measure pressures of this type are similar for both liquids and gases. For gases, however, it may be necessary also to measure pressures lower than atmospheric pressure, in some cases very much lower than atmospheric pressure. Such measurements are more specialised and employ quite different methods. We shall look first at the higher range of pressures in both gases and liquids.

The pressure sensors for atmospheric pressure or higher can make use of both indirect and direct effects. The indirect effects rely on the action of the pressure to cause displacement of a diaphragm, a piston or other device, so that an electronic measurement or sensing of the displacement will bear some relationship to the pressure. The best-known principle is that of the aneroid barometer, illustrated in Fig. 1.11. The diaphragm is acted on by the pressure that is to be measured on one side, and a constant (usually lower) pressure on the other side. In the domestic version of the barometer, the movement of the diaphragm is sensed by a system of levers which provide a pointer display of pressure.

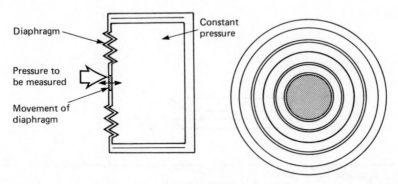

Fig. 1.11 The aneroid barometer principle. The domestic barometer uses an aneroid capsule with a low pressure inside the sealed capsule. Changes of external pressure cause the diaphragm to move, and in the domestic barometer these movements are amplified by a set of levers.

For electronic measurement, the diaphragm can act on any displacement transducer and one well-suited type is the capacitive type, illustrated in Fig. 1.12. The diaphragm is insulated from the fixed backplate, and the capacitance between the diaphragm and the backplate forms part of the resonant circuit of an oscillator. Reducing the spacing between the diaphragm and the backplate will increase the capacitance, in accordance with the formula shown in Fig. 1.12(b), and so reduce the resonant frequency of the oscillator. This provides a very sensitive detection system, and one which is fairly easy to calibrate. Though the thin metal corrugated diaphragm makes the device suitable only for detecting pressure around atmospheric pressure, the use of a thicker diaphragm, even a thick steel plate, can permit the method to be used with very much higher pressures. For such pressure levels, the sensor can be made in the form of a small plug that can be screwed or welded into a container. The smaller the cross-section of the plug the better when high pressures are to be sensed, since the absolute amount of force is the product of the pressure and the area of cross-section. The materials used for the pressure-sensing plate or diaphragm will also have to be chosen to suit the gas or liquid whose pressure is to be measured. For most purposes, stainless steel is suitable, but some very corrosive liquids or gases will require the use of more inert metals, even to the extent of using platinum or palladium.

Where a ferromagnetic diaphragm can be used, one very convenient sensing effect is variable reluctance, as illustrated in principle in Fig. 1.13. The

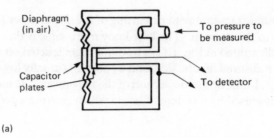

(a)

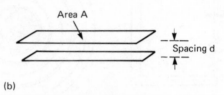

(b)

$$C = \frac{E_0 A}{d}$$
Units: E_0 – farads/metre
A – square metres
d – metres
C – farads

For air, $E_0 = 8.85 \times 10^{-12}$ F/m

Practical units:

$$C = \frac{8.85 \times 10^{-4} \times A}{d}$$
A – square centimetres
d – centimetres
C – picafarads

Fig. 1.12 The aneroid capsule arranged for pressure measurement. This is an inside-out arrangement as compared to the domestic barometer. The pressure to be measured is applied inside the capsule, with atmospheric air or some constant pressure applied outside. The movement of the diaphragm alters the capacitance between the diaphragm and a fixed plate, and this change of capacitance can be sensed electronically. The formula relating capacitance to spacing is shown in (b).

variable-reluctance type of pressure gauge is normally used for fairly large pressure differences, and obviously cannot be used where diaphragms of more inert material are required. The method can be used also for gases, and for a range of pressures either higher than atmospheric pressure or lower.

The aneroid barometer capsule is just one version of a manometer that uses the effect of pressure on elastic materials. Another very common form is the coiled flattened tube, as illustrated in Fig. 1.14, which responds to a change of pressure inside the tube (or outside it) by coiling or uncoiling. This type of sensor can be manufactured for various ranges of pressure simply by using

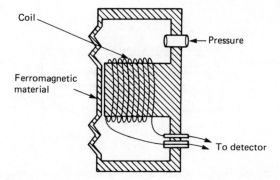

Fig. 1.13 Using a variable reluctance type of sensing system. The movement of the diaphragm causes considerable changes in the reluctance of the magnetic path, and so in the inductance of the coil.

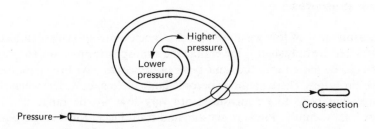

Fig. 1.14 The flattened-tube form of pressure sensor.

different materials and thicknesses of tubing, so that this method can be found used for both small and large pressure changes. The main drawback as far as electronics is concerned is the conversion from the coiling/uncoiling of the tube into electronic signals, and one common solution is to couple the manometer to a potentiometer.

Another transducing method is the use of a piezoelectric crystal, usually of barium titanate, to sense either displacement of a diaphragm or pressure on the crystal itself. As explained earlier, this is applicable more to short-duration changes than to steady quantities. For a very few gases, it may be possible to expose the piezoelectric crystal to the gas directly, so that the piezoelectric voltage is proportional to the pressure (change) on the crystal. For measurements on liquids and on corrosive gases, it is better to use indirect pressure, with a plate exposed to the pressure and transmitting pressure to the crystal, as in Fig. 1.15. This type of sensor has the advantage of being passive, with no need for a power supply to an oscillator and no complications of frequency measurement. Only a high input impedance voltmeter or operational amplifier is needed as an indicator, and if the sensor is used for switching purposes, the output from the crystal can be applied directly to a FET op-amp.

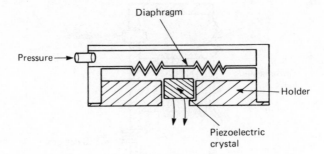

Fig. 1.15 Using a piezoelectric crystal detector coupled to a diaphragm for sensing pressure changes.

Low gas pressures

The measurement of low gas pressures is a much more specialised subject. Pressures that are only slightly lower than the atmospheric pressure of around 100 kPa can be sensed with the same types of devices as have been described for high pressures. These methods become quite useless, however, when the pressures that need to be measured are very low, in the range usually described as 'vacuum'. Pressure sensors and transducers for this range are more often known as vacuum gauges, and many are still calibrated in the older units of millimetres of mercury of pressure. The conversion is that 1 mm of mercury is equal to 133.3 Pa. The high-vacuum region is generally taken to mean pressures of 10^{-3} mm, of the order of 0.1 Pa, though methods for measuring vacuum pressures generally work in the region from about 1 mm (133.3 Pa) down. The most important of these methods are the Pirani gauge for the pressures in the region 1 mm to 10^{-3} mm (about 133 Pa to 0.13 Pa), and the ion gauge for significantly lower pressures down to about 10^{-9} mm, or 1.3×10^{-7} Pa.

The Pirani gauge, named for its inventor, uses the principle that the thermal conductivity of gases decreases in proportion to pressure for a wide range of low pressures. The gauge (Fig. 1.16) uses a hot wire element, and another wire as sensor. The temperature of the sensor wire is deduced from its resistance, and it is made part of a resistance measuring bridge circuit identical to that used for resistive strain gauges. As the gas pressure around the wires is lowered, less heat will be conducted through the gas, and so the temperature of the sensor wire will drop, since the amount of heat transmitted by convection is negligible (because of the arrangement of the wires) and the amount radiated is also very small because of the comparatively low temperature of the 'hot' wire. Commercially available Pirani gauges, such as those from Edwards High Vacuum or Leybold, are robust, easy to use, fairly accurate, and are not damaged if switched on at normal air pressures.

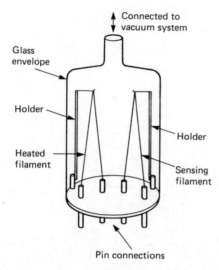

Fig. 1.16 The Pirani gauge. One filament is heated, and the other is used as a sensor of temperature by measuring its resistance. As the pressure in the air surrounding the filaments is decreased, the amount of heat conducted between the filaments drops, and the change in resistance of the cold filament is proportional to the change in pressure.

Ionisation gauges

For very low pressure, or high vacuum, measurement, some form of ionisation gauge is invariably used. There are many gauges of this type, but the principles are much the same and the differences are easily understood when the principles are grasped. The ionisation gauge operates by using a stream of electrons to ionise a sample of the remaining gas in the space in which the pressure is being measured. The positive gas ions are then attracted to a negatively charged electrode, and the amount of current carried by these ions is measured. Since the number of ions per unit volume depends on the number of atoms per unit volume, and this latter figure depends on pressure, the reading of ion current should be reasonably proportional to gas pressure. The proportionality is fairly constant for a fixed geometry of the gauge (Fig. 1.17) and for a constant level of electron emission. The range of the gauge is to about 10^{-7} mm (0.013 Pa), which is about the pressure used in pumping transmitting radio valves and specialised cathode ray tubes.

The most serious problem in using an ionisation gauge is that it requires electron emission into a space that is not a vacuum. The type of electron emitter that is used is invariably a tungsten filament, and if this is heated at any time when the gas pressure is too high (above about 10^{-3} mm (133 Pa)), then the filament will be affected. If, as is usual, the gas whose pressure is being reduced is air, the operation of the filament at these pressures will result in oxidation, which will impair electron emission or result in the total burnout

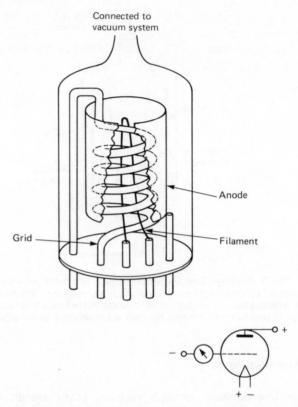

Connected to
vacuum system

Anode

Grid

Filament

Fig. 1.17 The simplest form of ionisation gauge. The grid is a loosely wound spiral of wire surrounding the filament, and exerts little control on the electron stream. With a constant high current of electrons to the anode, positive ions from the remaining gas are attracted to the grid and the resulting grid current is measured and taken as proportional to gas pressure.

of the filament. If ionisation gauges are used, as they nearly always are, in conjunction with other gauges, usually Pirani gauges, then it should be possible to interlock the supplies so that the ionisation gauge cannot be turned on until the pressure as indicated by the other gauge is sufficiently low. If this can be done, then the ionisation gauge can have a long and useful life. A spare gauge head should always be held in stock, however, in case of filament damage, because tungsten filaments are delicate, particularly when at full working temperature.

The variants on the ionisation gauge arise because a simple electron beam in a confined space is not necessarily a very efficient means of ionising the residual gas in that space, because only the atoms in the path of the beam can be affected. If the electron beam is taken through a longer path, more atoms can be bombarded, and more ions generated from a given volume of gas, and so the sensitivity of the device is greatly increased. The usual scheme is to use a magnetic field to convert the normal straight path of the electron beam into a spiral path which can be of a much greater total length. The very much greater sensitivity that can be obtained in this way is bought at the price of having another parameter, the magnetic field flux density, that will have to be controlled in order to ensure that correct calibration is maintained. The magnetic field is usually applied by means of a permanent magnet, so that day-to-day calibration is good, but since all permanent magnets lose field strength over a long period, the calibration should be checked annually. Gauges of this type can be used down to very low pressures, of the order of 10^{-9} mm (1.33^{-4} Pa).

Transducer use

The devices that have been described are predominantly used as sensors, because with a few exceptions their efficiency of conversion is very low and to achieve transducer use requires that the electrical signals be amplified. The piezoelectric device used for pressure sensing is also a useful transducer, and can be used in either direction. Transducer use of piezoelectric crystals is mainly confined to the conversion between pressure waves in a liquid or gas and electrical AC signals, and this use is described in detail in Chapter 5. The conversion of energy from electrical form into stress is achieved by the magnetically cored solenoid, as illustrated in Fig. 1.18. A current flowing in the coil creates a magnetic field, and the core will move so as to make the magnetic flux path as short as possible. The amount of force can be large, so that stress can be exerted (causing strain) on a solid material. If the core of the solenoid is mechanically connected to a diaphragm, then the force exerted by the core can be used to apply pressure to a gas or a liquid. In general, though, there are few applications for electronic transducers for strain or pressure and the predominant use of devices in this class is as sensors.

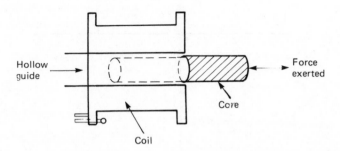

Fig. 1.18 The solenoid, which is a current to mechanical stress transducer.

Chapter Two
Position, Direction, Distance and Motion

Position

Position, as applied in measurement, invariably means position relative to some point which may be the Earth's north pole, the starting point of the motion of an object, or any other convenient reference point. Methods of determining position make use of distance and direction (angle) information, so that a position can be specified either by using rectangular (Cartesian) coordinates (Fig. 2.1) or by polar coordinates (Fig. 2.2). Position on flat

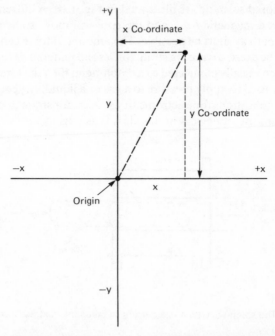

Fig. 2.1 The Cartesian coordinate system. This uses two directions at right angles to each other as reference axes, and the position of a point is plotted by finding its distance from each axis. For three-dimensional location, three axes, labelled x, y and z, can be used. The figure also shows conversion of two-dimensional Cartesian coordinates to polar form.

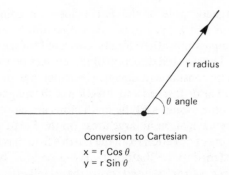

Conversion to Cartesian

$x = r \cos \theta$
$y = r \sin \theta$

Fig. 2.2 Polar coordinates make use of a fixed point and direction. The distance from the fixed point, and the angle between this line and the fixed direction, are used to establish (two-dimensional) position. For three-dimensional location, an additional angle is used. The figure also shows conversion of two-dimensional polar coordinates to Cartesian.

surfaces, or even on the surface of the Earth, can be specified using two dimensions, but for air navigational purposes three-dimensional coordinates are required. For industrial purposes, positions are usually confined within a small space (the position of a robot tug, for example) and it may be possible to specify position with a single number such as the distance travelled along a rail.

In this chapter we shall look at the methods that are used to measure direction and distance so that position can be established either for large or small-scale ranges of movement. There are two types of distance sensing – the sensing of distance to some fixed point, and the sensing of distance moved – which are different both in principle and in the methods that have to be used. The methods that are applied for small-scale sensing of position appear at first glance to be very different, but they are in fact very similar in principle. Since position is related to distance (the difference between two positions), velocity (rate of change of position) and acceleration (rate of change of velocity), we shall look at sensors for these quantities also. Rotational movement is also included because it is very often the only movement in a system and requires rather different methods. In addition, of course, the rotation of a wheel is often a useful measurement of distance moved.

Direction

The sensing of direction on the Earth's surface can be achieved by using the Earth's magnetic field, by making use of the properties of gyroscopes, or by radio methods, the most modern of which are satellite direction-finders. Starting with the most ancient method, the traditional compass uses the effect of the Earth's magnetic field on a small magnetised needle which is freely suspended so that the needle points along the line of the field, in the direction of magnetic north and south. The qualifying word 'magnetic' is important here. The north magnetic pole of the Earth does not coincide with the geographical pole, nor is it a fixed point. Any direction found by use of a magnetic form of compass must therefore be corrected for true north if high accuracy is required. The size and direction of this correction can be obtained from tables of magnetic constants (magnetic elements) that are published for the use of navigators. The drift speed and direction of the magnetic north pole can be predicted to some extent, and the predictions are close enough to be useful in fairly precise navigation in large areas on the Earth's surface.

For electronic sensing of direction from the Earth's field, it is possible to use a magnetic needle fastened to the shaft of a servo-generator, but this type of mechanical transducer is rarely used now that Hall-effect sensors are available. The Hall effect is an example of the action of a magnetic effect on moving charged particles, such as electrons or holes, and it was the way in which hole movement in metals and semiconductors was first proved. The principle is a comparatively simple one, but for most materials detecting the effect requires very precise measurements.

The principle is illustrated in Fig. 2.3. If we imagine a slab of material carrying current from left to right, this current, if carried entirely by electrons, would consist of a flow of electrons from right to left. Now for a current and a field in the directions shown, the force on the conductor will be upwards,

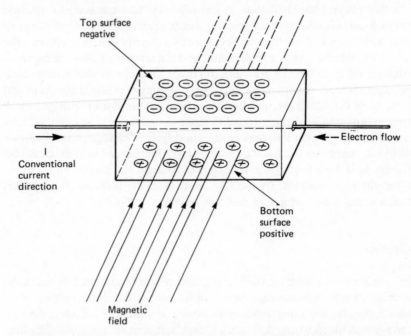

Fig. 2.3 The Hall effect. Hall showed that the force of a magnetic field on a current carrier was exerted on the carriers, and would cause deflection. The deflection leads to a difference in voltage across the material, which is very small for a metal because of the high speeds of the carriers, but much larger for a semiconductor.

and this force is exerted on the particles that carry the current, the electrons. There should therefore be more electrons on the top surface than on the bottom surface, causing a voltage difference, the Hall voltage, between the top and bottom of the slab. Since the electrons are negatively charged, the top of the slab is negative and the bottom positive. If the main carriers are holes, the voltage direction is reversed.

The Hall voltage is very small in good conductors, because the particles move so rapidly that there is not enough time to deflect a substantial number in this way unless a very large magnetic field is used. In semiconductor materials, however, the particles move more slowly, and the Hall voltages can be quite substantial, enough to produce an easily measurable voltage for only small magnetic fields such as the horizontal component of the Earth's field. Small slabs of semiconductor are used for the measurement of magnetic fields in Hall-effect fluxmeters and in electronic compasses. A constant current is passed through the slab, and the voltage between the faces is set to zero in the absence of a magnetic field. With a field present, the voltage is proportional to the size of the field, but the practical difficulty is in determining direction. The direction of maximum field strength is in a line drawn between the magnetic north and south poles, but because the Earth is (reasonably exactly) a sphere, such a line, except at the equator, usually is directed into the Earth's surface, and the angle to the horizontal is known as the angle of dip (Fig. 2.4). The conventional magnetic compass needle gets around this problem by being pivoted and held so that it can move only in a horizontal plane, and this is also the solution for the Hall-effect detector. A precision electronic compass uses a servomotor to rotate the Hall slab under the control of a discriminator circuit which will halt the servomotor in the direction of maximum field strength with one face of the Hall slab positive. By using an analogue to digital converter for angular rotation, the direction can be read out in degrees, minutes and seconds. The advantages of this system are that the effects of bearing friction that plague a conventional compass are eliminated, and the reading is not dependent on a human estimate of where a needle is placed relative to a scale.

The global nature of the Earth's magnetic field makes it particularly convenient for direction sensing, but the irregular variations in the field cause problems, and other methods are needed for more precise direction finding, particularly over small regions. The properties of a gyroscope have been used for this purpose for a considerable time, and gyrocompasses were developed to a very considerable degree for air navigation in World War II. The principle is that a spinning flywheel has directional inertia, meaning that it resists any attempt to alter the direction of its axis. If the flywheel is suspended so that the framework around it can move in any direction without exerting force on the flywheel, then if the axis of the flywheel has been set in a known position, such as true north, this direction will be maintained for as long as the flywheel spins. Suitable suspension frameworks have been developed from the old-fashioned gimbals which are used for ships compasses, and the

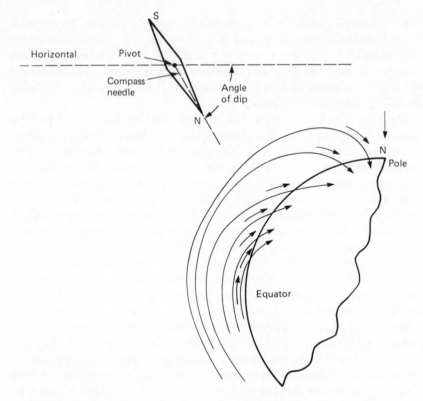

Fig. 2.4 The angle of dip shows the actual direction of the Earth's field, which in the northern hemisphere is always into the surface of the Earth.

wartime gyrocompasses maintained the rotation of the spinning wheel by means of compressed air jets.

The gyrocompass has no inherent electrical output, however, and it is not a simple matter to obtain an electrical output without placing any loading on the gyro wheel. Laser gyroscopes making use of rotating light beams have been developed, but are extremely specialised and beyond the scope of this book. In addition, gyroscopes are not used to any extent in small-scale direction finding for industrial applications.

Radio has been used for navigational purposes for a long time, in the form of radio beacons that are used in much the same way as light beacons were used in the past. The classical method of using a radio beacon is illustrated in Fig. 2.5, and consists of a receiver that can accept inputs from two aerials, one a circular coil that can be rotated and the other a vertical whip. The signal from the coil aerial is at maximum when the axis of the coil is in line with the transmitter, and the phase of this maximum signal will be either in phase with the signal from the vertical whip aerial or in antiphase, depending on whether the beacon transmitter is ahead or astern of the coil. By using a phase-sensitive receiver that indicates when the phases are identical, the position of maximum

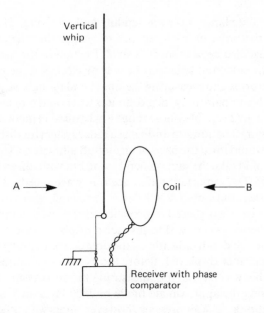

Fig. 2.5 The radio direction finder principle. The output from the vertical aerial is obtained from the electrostatic field of the wave, and does not depend on direction. The magnetic portion of the wave will induce signals in a coil, but the phase of these signals depends on the direction of the transmitter. By combining the signals from the two aerials, and turning the coil, the direction of the transmitter can be found as the direction of maximum signal.

signal ahead can be found, and this will be the direction of the radio beacon.

Satellite direction-finding is an extension of these older systems and depends on the supply of geostationary satellites. A geostationary satellite is one whose angular rotation is identical to that of the Earth, so that as the Earth rotates the satellite is always in the same position relative to the surface of the planet. The navigation satellites are equipped with transponders which will re-radiate a coded received signal. At the surface, a vessel can send out a suitably coded signal and measure the time needed for the response. By signalling to two satellites in different positions, the position on the Earth's surface can be established very precisely – the precision depends on the frequency that is used, and this is generally in the millimetre range.

Distance measurement – large scale

The predominant method of measuring distance to a target point on a large scale is based on wave reflection of the type used in radar or sonar. The principle is that a pulse of a few waves is sent out from a transmitter, reflected at some distant object and detected by a receiver when it returns. Since the speed of the waves is known, the distance of the reflector can be calculated from the time that elapses between sending and receiving. This time can be very short, of the order of microseconds or less, so that the duration of the wave pulse must also be very short, a small fraction of the time that is to be measured. Both radar and sonar rely heavily on electronic methods for generating the waveforms and measuring the times, and though we generally associate radar with comparatively long distances we should remember that radar intruder alarms are available whose range is measured in metres rather than in kilometres. Figure 2.6 shows an outline of a radar system for distance measurement, such as would form the basis of an aircraft altimeter; a sonar system for water depth would take the same general form, but with different transducers (see Chapter 5). The important difference is in wave speeds, 3×10^8 m/s for radio waves in air, but only 1.5×10^3 m/s for sound waves in sea-water.

Where radar or sonar is used to provide target movement indications, the time measurements will be used to provide a display on a cathode ray tube, but for altimeters or depth indications, the time can be digitally measured and the figure for distance displayed. Before the use of radar altimeters, the only method available was barometric, measuring the air pressure by an aneroid capsule and using the approximate figure of 3800 Pa change of pressure per kilometre of altitude. The air pressure, however, alters with other factors such

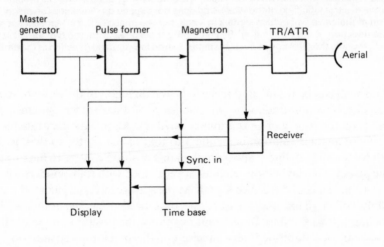

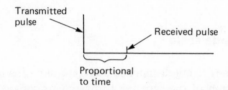

Fig. 2.6 The block diagram for a simple radar system. The time required for a pulse of microwave signal to travel to the target and back is displayed as a distance on a cathode ray tube. The transmitter and receiver share the same aerial, using a TR/ATR (Transmit/ Antitransmit) stage to short circuit the receiver while the transmitted pulse is present.

as humidity, wind-speed and temperature, so that pressure altimeters are notoriously unreliable. Even if such an altimeter were to give a precise reading, the height that it measures will either be height above sea level or the height relative to the altitude of the place in which the altimeter was set, rather than true height. It is, in fact, remarkable that air travel ever became a reality with such a crude method of height measurement.

Position measurement on a smaller scale (factory floor scale) can make use of simpler methods, particularly if the movement is confined in some way, such as by rails or by the popular method of making a robot trolley follow buried wires. For confined motions on rails or over wires, distance from a starting point may be the only measurement that is needed, but it is more likely that movement is two-dimensional. Over small areas of a few square metres, an artificially generated magnetic field can be used along with magnetic sensors of the types already described. Radio beacon methods, using very low power transmitters, are also useful, and ultrasonic beacons can be used, though problems arise if there are strong reflections from hard surfaces. For a full discussion of the methods as distinct from the sensors, the reader should consult a text on robotics.

Distance travelled

The sensing of distance travelled as distinct from distance from a fixed reference point can make use of a variety of sensors. In this case, we shall start with the sensors for short distance movements, because for motion over large distances the distance travelled will generally be calculated by comparing position measurements rather than directly. Sensors for small distances can make use of resistive, capacitive or inductive transducers in addition to the use of interferometers (see Chapter 1) and millimetre-wave radar methods that have been covered earlier. The methods described here are all applicable to distances in the range of a few millimetres to a few centimetres; beyond this range the use of radar methods becomes much more attractive.

A simple system of distance sensing is the use of a linear (in the mechanical sense) potentiometer (Fig. 2.7). The moving object is connected to the slider of the potentiometer, so that each position along the axis will correspond to a different output from the slider contact – either AC or DC can be used since only amplitude needs to be measured. The output can be displayed on a meter, converted to digital signals to operate a counter or used in conjunction with voltage level sensing circuits to trigger some action when the object reaches some set position. The main objections to this potentiometric method are that the range of movement is limited by the size of potentiometers that are available (though purpose-built potentiometers can be used) and that the friction of the potentiometer is an obstacle to the movement. The precision that can be obtained depends on how linear (in the electrical sense) the winding can be made, and 0.1% should be obtainable with reasonable ease.

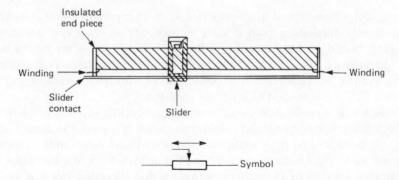

Fig. 2.7 A sensor for linear displacement in the form of a linear potentiometer. The advantage of this type of sensor is that the output can be a steady DC or AC voltage which changes when the displacement changes.

An alternative that is sometimes more attractive but often less practical is the use of a capacitive sensor. This can take the form of a metal plate located on the moving object and moving between two fixed plates that are electrically isolated from it. The type of circuit arrangement is illustrated in Fig. 2.8, showing that the fixed plates are connected to a transformer winding so that AC signals in opposite phase can be applied. The signal at the moveable plate will then have a phase and amplitude that depends on its position, and this signal can be processed by a phase-sensitive detector to give a DC voltage that is proportional to the distance from one fixed plate. Because the capacitance between plates is inversely proportional to plate spacing, this method is practicable only for very short distances, and is at its most useful for distances of a millimetre or less. An alternative physical

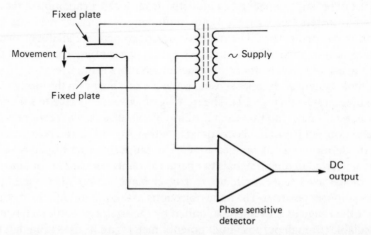

Fig. 2.8 The capacitor plate sensor in one of its forms. A change in the position of the moving plate will cause the voltage between this plate and the centre tap of the transformer to change phase, and this phase change can be converted into a DC output from the phase-sensitive detector.

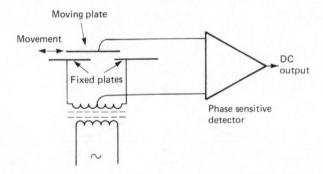

Fig. 2.9 Another form of the three-plate capacitor displacement detector. In this case, the spacing between the plates is fixed, but the relative area of centre plate covering the other two can vary. The range of movement for this type can be considerably greater than for the previous type.

arrangement of the plates is shown in Fig. 2.9, in which the spacing of the fixed plates relative to the moving plate is small and constant, but the movement of the moving plate alters the area that is common to the moving plate and a fixed plate. This method has the advantage that an insulator can be used between the moving plate and the fixed plates, and that the measurable distances can be greater, since the sensitivity depends on the plate areas rather than on variable spacings.

The most commonly used methods for sensing distance travelled on the small scale, however, depend on induction. The basic principle of induction methods is illustrated in Fig. 2.10, in which two fixed coils enclose a moving ferromagnetic core. If one coil is supplied with an AC signal, then the amplitude and phase of a signal from the second coil depends on the position of the ferromagnetic core relative to the coils. The amplitude of signal, plotted against distance from one coil, varies as shown in Fig. 2.11, and the disadvantage of this simple arrangement is that a given amplitude other than the maximum can correspond to more than one distance. In addition, the shape of the graph means that even if the range is restricted the output is never linearly proportional to the distance.

A development of the simple inductive sensor is the linear variable differential transformer (LVDT), which is now the most commonly used

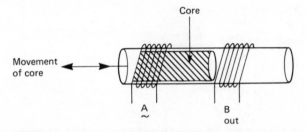

Fig. 2.10 The most basic inductive displacement sensor. An AC voltage is applied to one coil, and the position of the core determines how much will be picked up by the other coil.

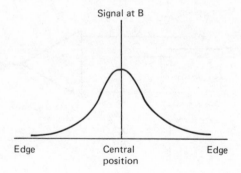

Fig. 2.11 A graph of output voltage plotted against core position for the arrangement of Fig.2.10.

sensor for distance in the range of millimetres to centimetres. The principle is illustrated by the circuit diagram of Fig. 2.12. The device consists basically of three fixed coils, one of which is energised with AC. The other two coils are connected to a phase-sensitive detector, and as a core of ferromagnetic material moves in the coil axis, the output from the detector will be proportional to the distance of the core from one end of the coils. As the name suggests, the output from the phase-sensitive detector will be fairly linearly proportional to distance, and there are considerable advantages as compared to other types of distance sensors. These are:

(1) Virtually zero friction, since the core need not be in contact with the coils, and so no wear.
(2) Linear output.
(3) Very high resolution, depending mainly on the detector.
(4) Good electrical isolation between the core and the coils.

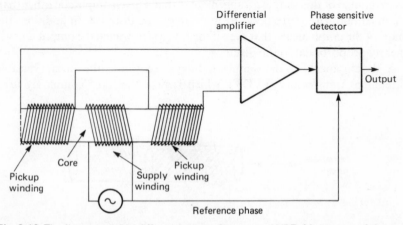

Fig. 2.12 The linear variable differential transformer, or LVDT. Movement of the core alters the voltage levels and phases of the voltages across the outer coils, and these voltages can be converted into DC by the phase-sensitive detector.

(5) A large output signal from the coils so that the phase-sensitive detector needs little or no amplification.
(6) No risk of damage if the core movement is excessive.
(7) Strong construction that is resistant to shock and vibration.

Because of these advantages, the LVDT has superseded most other types of distance sensors. For very small distances, the strain gauge (see Chapter 2) can be used; laser interferometers are applicable when very precise changes must be sensed, and radar methods are used for longer distances.

One peculiar advantage of the laser interferometer is that its output can readily be converted to digital form, since it is based on the counting of wave-peaks. Most sensors give analogue outputs, and where a sensor is described as having a digital output, this usually implies that an analogue to digital conversion has been carried out. The interferometer is one of the few sensors which is capable of providing a genuinely digital output. Another is the linear digital encoder, and since digital methods are of ever-increasing importance in measurement and control, a description of this device in detail is appropriate here.

The linear digital encoder gives an output which is a binary number that can be proportional to the distance of the encoder relative to a fixed point. The encoder can, of course, be fixed with an object moving relative to it. The simplest form of encoder of this type is optical, and the principle is illustrated in Fig. 2.13. A glass strip is printed with a pattern of the form shown, using large blocks rather than the more practical lines, alternately opaque and transparent. At one edge, the pattern is comparatively fine, and the size of each unit determines the resolution of the device – the smallest change of distance that can be sensed is the width of one block or bar. The next strip along contains blocks or bars of double the width of the first, again alternately opaque and transparent. The next strip in turn has bars that are twice the width of its predecessor and so on. The number of strips determines the number of binary digits in the output and the number of increments of position that can be detected. For example, if the pattern consists of four strips, then the maximum number of increments of position is $2^4=16$ – not likely to be of practical use except for specialised purposes. For eight strips, however, the number of resolvable position is $2^8=256$, and for 16 strips, the number is 65 536.

The optical linear encoder is read by using a sensor for each strip or track. The usual scheme is to use a strip of photocells so that light from a source on the other side of the encoder strip will pass through the transparent sections to the appropriate photocell. The output of each photocell will therefore, after amplification, be a 1 or a 0 signal, and the set of photocells provides a binary number if the cells are read in order of the broadest bars towards the finest. If an eight bit number is used, with eight strips and eight photocells, then the number range is 0 to 256, and this determines the resolution as being 1/256 of the total length of the encoder. For example, if the encoder is 10 cm long,

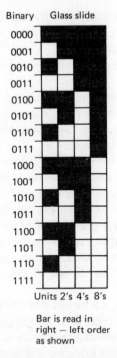

Binary   Glass slide

0000
0001
0010
0011
0100
0101
0110
0111
1000
1001
1010
1011
1100
1101
1110
1111

Units 2's 4's 8's

Bar is read in
right — left order
as shown

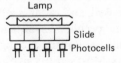

Lamp

Slide

Photocells

Fig. 2.13 The binary optical encoder. Moving the glass strip between a light source and a set of photocells will result in outputs from each photocell which will make up a binary number. The main problem of using 8-4-2-1 binary code with this system is that incorrect readings can be obtained when the position of the slide is such that sections overlap the photocells. The resolution is equal to the width of the units mark.

then the resolution is 10/256 cm, or 0.39 mm. The device can be used with the light and photocells at rest and the encoder slide moving (the normal use), or with the encoder slide at rest and the light source and/or photocells moving. The photocells must be placed almost in contact with the encoder slide, and with suitable light shielding to ensure that stray light does not cause false responses.

One problem that arises with optical linear (or angular) encoders is the suitability of the coding system. The usual binary number system is often termed 8-4-2-1 code, in recognition of the fact that each digit place represents a doubling or halving of the value of the adjacent place. This can mean that some changes of number can involve large changes in the digits. For example, the change from 7 to 8 in binary terms is the change from 0111 to 1000, in which each of four digits has changed simply to indicate a value change of just one unit. For this reason, another code, the Grey code (illustrated in Fig. 2.14), is used to a much greater extent for devices such as optical encoders. In a Grey code, a change of one unit will affect only one digit of the binary number, so that the chances of error caused by misalignment of the sectors of the coding glass, or by half-way positions between sectors, are greatly

Denary	Grey Code	8-4-2-1 Binary
0	0000	0000
1	0001	0001
2	0011	0010
3	0010	0011
4	0110	0100
5	0111	0101
6	0101	0110
7	0100	0111
8	1100	1000
9	1101	1001
10	1111	1010
11	1110	1011
12	1010	1100
13	1011	1101
14	1001	1110
15	1000	1111

Fig. 2.14 The Grey scale, which is widely used for optical encoders and for industrial purposes generally. Unlike 8-4-2-1 binary, the Grey code alters by only one digit for each unit change, so that errors due to positioning a slide so that a section affects more than one photocell are negligible.

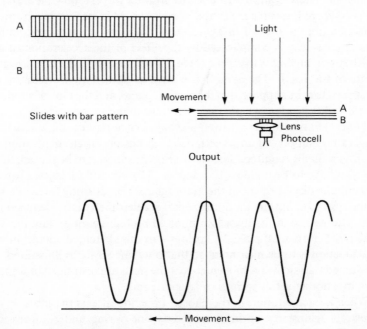

Fig. 2.15 The moiré fringe detector. The two slides each have identical thin line patterns printed on them. When one slide moves relative to the other, the amount of light passing through varies considerably depending on the relative positions of lines and spaces in the two slides. A photocell can detect this output and count peaks to find a total displacement.

reduced. ICs exist which will convert Grey code into normal 8-4-2-1 binary, so that arithmetic can be carried out on the numbers if needed. For some applications, conversion may not be necessary.

Another optical method that is particularly useful for small displacements, particularly vibration amplitudes, is the optical grating method whose principles are illustrated in Fig. 2.15. Each grating consists of a glass or plastic plate on which fine lines are engraved. One grating is fixed and the other is part of the object whose movement is to be detected. As the object moves, the amount of light that passes through both gratings will be altered – ideally in the pattern of a sine wave for a movement that is of an amplitude equal to the distance between grating lines. This alteration of the light amplitude can be detected by a photocell and used to provide an electrical output. Large movements are read in terms of the number of complete waves of output from the photocell, small movements in terms of the amplitude of the signal.

Accelerometer systems

For sensing the quantities of acceleration, velocity and distance travelled, systems based on accelerometers are used. The basis of all accelerometers is the action of acceleration on a mass to produce force, following the equation $F = Ma$ where F is force measured in newtons, M is mass in kilograms and a is acceleration in units of m/s^2. The use of a mass, often termed an inertial mass, in this way is complicated by the effect of the acceleration of gravity which causes any mass to exert a force (its weight) that is directed towards the centre of the Earth. The mass that is used as part of an acceleration sensor must therefore be supported in the vertical plane, and the type of support that is used will depend on whether accelerations in this plane are to be measured. If the acceleration to be measured is always in a horizontal direction, then the mass can be supported on wheels, ball bearings or air-jets, depending on the sensitivity that is required. Since a force on the mass is to be sensed, the mass will also have to be coupled to a sensor. The method of sensing force is to measure the displacement of the mass against the restoring force of a spring, so the system in outline for a horizontal accelerometer is as sketched in Fig. 2.16. The weight of the mass is supported on ball bearings, and the mass is held in the horizontal plane by springs – in this sketch, acceleration in only one direction is to be measured, so that only one spring is illustrated, and it is assumed that sideways movements of the mass are restrained in some other way, by another set of springs or by guide rails.

When an acceleration in the chosen direction affects the mass, it will be displaced against the springs, extending one spring and compressing the other. The amount of linear movement will be proportional to the force, so that any sensor for linear displacement can be used to give an output that is proportional to acceleration. Suitable transducers include potentiometers,

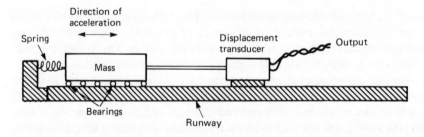

Fig. 2.16 Measuring acceleration in one direction. The acceleration of the mass causes a force equal to mass × acceleration. This is turn will stretch or compress the spring, and the amount of this displacement can be measured by any of the usual methods, usually LVDT.

capacitive distance gauges, inductive gauges and LVDTs for the larger ranges of movement that are possible. Sensors for acceleration that use this spring and displacement principle are generally intended for the measurement of very small accelerations, usually in one plane. If the mass can be supported in a cradle of springs and three displacement sensors connected, one for each axis of motion (two horizontal and one vertical), then the outputs can be used to compute the magnitude and direction of an acceleration that can be in any direction. If only a single dimensional acceleration can be measured, the acceleration in any other direction will produce a false reading, equal to the component of acceleration in that direction. Figure 2.17 shows how this component of acceleration is related to the true value and to the angle between the true acceleration and the measured acceleration directions.

A strain-gauge sensor can be connected to an inertial mass, with or without spring suspension in order to measure acceleration, but the most common type of accelerometer uses an inertial mass coupled to a piezoelectric crystal. The piezoelectric effect has been known since the end of the nineteenth century, though even in 1960 one major University textbook of physics did not mention it. The principle is that some crystalline materials such as quartz (silicon oxide), Rochelle salt and barium titanate are composed of charged particles (ions) which do not move uniformly when the crystal is stressed. Because of this non-uniformity, movement of the ions produces a difference in charge between opposing faces of the crystal, and if these faces are

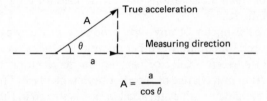

Fig. 2.17 The relationship between acceleration measured out of line and true acceleration, if the correct direction is known.

metallised, a voltage can be measured. The voltage is proportional to the strain of the crystal and can be very large, of the order of kilovolts for some ceramic crystal types, if the strain of the crystal is large. This principle is used to provide a spark for cigarette lighters and gas ignition systems.

The use of an inertial mass bonded to a piezoelectric crystal therefore provides an accelerometer that requires no springs or special supports for the mass. It is even possible to obtain two-dimensional signals from one crystal, and the system will respond to a very wide range of accelerations. Though the unit of acceleration is the m/s^2, acceleration is very often measured relative to the 'standard' value of the acceleration of gravity, which is 9.81 m/s^2. This leads to figures of acceleration in 'g' units, from which the value in scientific units can be calculated by multiplying by 9.81. Piezoelectric transducers can cope with acceleration values from a very small fraction of 'g' to several thousand 'g', a huge range compared to those that can be obtained by using spring and displacement systems. The snag is that the piezoelectric crystal is, from the circuit point of view, a capacitor, and the signal is in the form of a charge. The connection of a resistance to the contacts on the faces of the crystal will therefore allow this capacitance to discharge with a time constant equal to CR seconds, where C is the capacitance between the crystal faces in μF and R is the resistance between the faces in Mohms. If, for example, the capacitance is 1000 pF (= 0.001 μF) and the resistance is 10 000 M (the input of a FET DC amplifier, perhaps), then the time constant is only 10 seconds. This makes the piezoelectric type of sensor more suitable for measuring changes of acceleration that occur over a short time, perhaps a fraction of a second, than for measuring fairly constant values of acceleration for a long period.

As it happens, a lot of accelerations are not sustained. Newton's first law states in effect that the natural state of any object in the universe is uniform motion in a straight line with no form of acceleration, and acceleration comes about only because of force. The only steady and constant value of acceleration that we encounter is the acceleration of gravity, and most of our acceleration measurements are on accelerations that are caused by short-duration forces, such as those encountered when one object hits another. To put this into perspective, an acceleration of only 1 g for 10 seconds corresponds to falling a distance of about 500 metres in a vacuum. In general, unless you are working with propulsion systems for outer space, the measurement of small accelerations that are applied for long periods will be of no practical interest.

When an accelerometer of any type produces an electrical output, this output can be used for computing other quantities. One of these quantities is speed or, more correctly, velocity because an acceleration exists when a change of direction or a change of speed or both take place. The relationship between speed or velocity value and acceleration is shown in Fig. 2.18, so that if the starting speed of an object, its acceleration (assumed constant) and the time of acceleration are all known, then the final speed can be calculated. The

For uniform (steady) acceleration:

$$\text{acceleration} = \frac{\text{change of speed}}{\text{time taken}} \quad \text{(direction unchanged)}$$

For non-uniform acceleration:

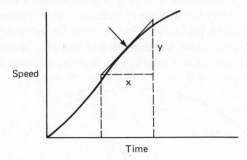

Acceleration value at arrowed point $= \dfrac{y}{x}$

Note: change of direction also constitutes acceleration. For circular motion, the revolving object has acceleration equal to $\dfrac{v^2}{r}$, directed to the centre of rotation, where v = linear velocity at any instant, r = radius of circle.

Fig. 2.18 The relationship between acceleration and velocity (speed) if the direction is constant.

mathematical action needed to find change of speed from acceleration and time is called integration, and analogue computers can carry out this action on a voltage signal from an accelerometer. The initial speed can be set in the form of a voltage applied from a potentiometer, and the output of the analogue computer is proportional to final speed. A second integration of the voltage output (the speed output) will produce a signal proportional to distance, so that this quantity also can be found by using an analogue computing action on the output of the accelerometer. If the starting point of the motion is rest (no starting speed), then no constants need to be fed in. The analogue computer can consist of little more than a pair of operational amplifiers if a simple one-dimensional motion is being sensed.

Rotation

There are very few machines that do not include a rotating shaft at some place, and the sensing and measurement of rotational movement is therefore important. The quantities that are used to measure rotation correspond to the quantities that are used in the measurement of linear motion in such a way that the same types of equations can be used, substituting the rotational quantities for the linear ones. The quantity that corresponds to distance for a rotation is the angle rotated. For a complete rotation of a shaft, this angle is 360 degrees, and so the total angle turned by a shaft is 360 × the number of complete turns, in terms of degrees. The degree, however, is an artificial unit that is not used in calculations, and most textbooks show the relationships between rotational quantities in terms of radians. The definition of the radian is illustrated in Fig. 2.19, and it leads to the angle for one complete rotation being 2π radians. To convert from degrees to radians, divide the angle in degrees by 57.3; to convert from radians to degrees, multiply by 57.3.

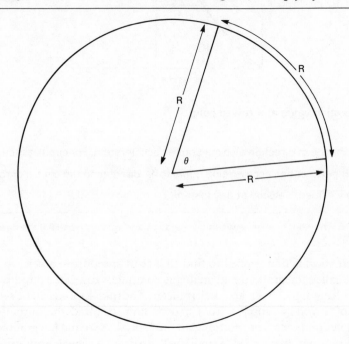

The angle θ is 1 radian, for any size of circle.

Relationship with degrees:

Complete circumference of circle = 2πR for 360 degrees

Since 1 radian corresponds to R distance, the radian

is $\dfrac{360°}{2\pi} = \dfrac{180°}{\pi} = 57.296°$

and $1° = \dfrac{\pi}{180}$ rad. $= 0.01745$ rad.

Fig. 2.19 The radian, and how it relates to degree measure of angles.

Because rotational angle corresponds to linear distance, rotational speed is defined as the angle through which a shaft turns per second, and corresponds to linear speed. If the rotating object is a wheel and is in contact with a surface, then Fig. 2.20 shows how the linear distance and velocity are related to the angle turned and the angular velocity. The angular acceleration is defined as the rate of change of angular velocity, and the equations that relate these angular quantities, along with the linear counterparts, are illustrated in Fig. 2.21. The sensors for angular motion are also very closely related to those that are used for linear motion.

Taking angular velocity first, the simplest form of sensor, which can also

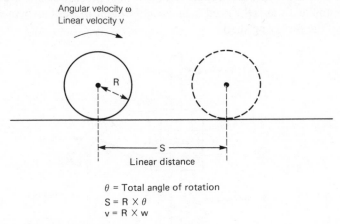

Angular velocity ω
Linear velocity v

S
Linear distance

θ = Total angle of rotation
$S = R \times \theta$
$v = R \times w$

Fig. 2.20 Linear and angular distance and speed.

Linear	Angular
$v = u + at$	$\omega_2 = \omega_1 + \alpha t$
$s = ut + \frac{1}{2}at^2$	$\theta = \omega t + \frac{1}{2}\alpha t^2$
$v^2 = u^2 + 2as$	$\omega_2^2 = \omega_1^2 + 2\alpha\theta$
$F = ma$	$\Gamma = I\alpha$
Momentum = mv	Angular momentum = Iw
Energy = $\frac{1}{2}mv^2$	Energy = $\frac{1}{2}I\omega^2$

u, v	velocity	ω_1, ω_2	angular velocity
s	distance	θ	angle
F	force	Γ	torque
m	mass	I	moment of inertia
a	acceleration	α	angular acceleration
t	time	t	time

Fig. 2.21 The relationships between linear and angular quantities illustrated by corresponding equations.

act as a transducer, is the AC or DC generator. For sensing and measurement purposes only a minimum of power must be used, so that a miniature AC generator called the tacho-generator is normally used. The construction of the tacho-generator, usually abbreviated to tacho, is a more precision-built version of any AC generator and usually has rotating magnets with output from stator coils so as to avoid the need for slip-rings. The frequency of the output signal is proportional to the revolutions per second of the shaft that is coupled to the tacho, so that a frequency-sensitive detector can be used to give a DC output proportional to angular frequency. The tacho can be used over a wide range of angular velocity values, and if the frequency detector is reasonably linear then the readings can be of high precision. The drawback for some applications is the need to make a mechanical coupling between the tacho and the revolving shaft.

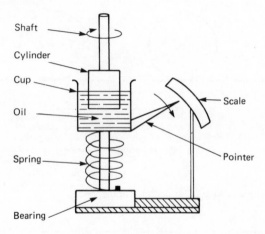

Fig.2.22 The principle of the drag cup method of measuring angular velocity. The method can be converted to electrical output by dispensing with the pointer and mounting the cup and its retaining spring on to the shaft of a potentiometer.

Another form of measurement depends on 'drag systems'. A drag system involves a frictional coupling between the rotating shaft and some object that is restrained and whose displacement can be measured. One version of this is the drag cup, whose principle is illustrated in Fig. 2.22. The end of the vertical rotating shaft dips into oil contained in a cup, and the cup is held in a spring mounting so that its rotation can be measured by any type of sensor, which can be a potentiometer, capacitor system, rotary LVDT or digital encoder (see later). The motion of the shaft is communicated by the viscosity of the oil to a turning force (torque) on the cup, and the displacement of the cup against the spring is measured by the sensor. Since displacement should be proportional to torque, which in turn should be proportional to rotational speed of the shaft, output from the sensor is proportional to rotational speed over a small range of speeds. The range is small because the assumption that torque is proportional to rotational speed holds good for only a small range of speeds, and the system is best suited for slow rotations. Another version of this method which is much more versatile is the magnetic disc type. A magnet on the end of the shaft will cause a rotating field as the shaft turns, and if a metal disc (which need not be magnetic) is held close to this magnet (Fig. 2.23) then the torque on the disc (caused by the interaction of the magnet and the eddy currents that are induced in the disc) will be proportional to the angular speed of the shaft. This is the scheme which has been used for many years for car speedometer heads, and it operates best at a medium range of angular speeds. One particular advantage is that no contact is needed, though there must not be any metal between the magnet and the disc.

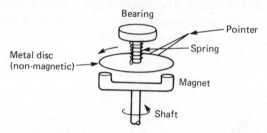

Fig. 2.23 The principle of magnetic drag, using induced currents. This has been extensively used in speedometers, and can be adapted to electronic measurement by replacing the disc and its mountings by a sensing coil.

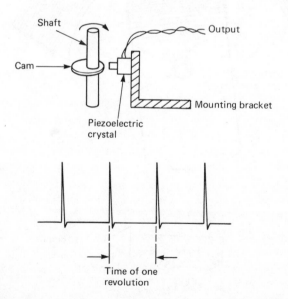

Fig. 2.24 Rotational speed sensing using a cam and a piezoelectric crystal. The disadvantages are that there is only one pulse per revolution, and that the contact between cam and crystal causes a frictional drag on the rotation of the shaft.

For some purposes, a signal that is sent out for each revolution of a wheel or shaft is sufficient for angular velocity sensing, and this can be achieved by the use of piezoelectric or magnetic pulsing. A piezoelectric pulse can be operated, as indicated in Fig. 2.24, by a cam on a shaft which will cause the piezoelectric crystal to be compressed on each rotation of the shaft. Since the signal from the piezoelectric crystal can be of several volts amplitude, this type of sensor often needs no amplification, but the output is at a high impedance. In addition, the friction on the shaft is fairly large compared to the alternative system of magnetic pulsing. The magnetic pulse system uses a permanent magnet mounted on a wheel or shaft and passing over a coil at one part of the revolution. As the magnet moves, a voltage is induced in the coil, so providing a signal for each revolution. The magnetic pulse system is not ideal if the rotation is slow, because the amplitude of the pulse is affected by the rotational speed, and there must not be any magnetic metal between the magnet and the coil.

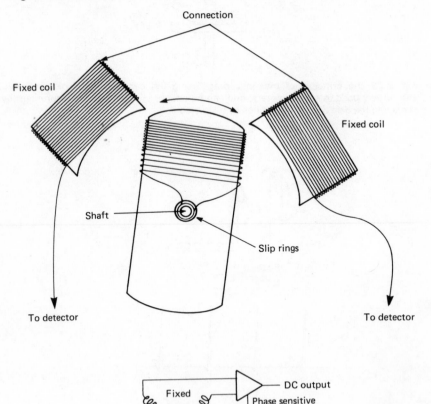

Fig. 2.25 A rotary form of LVDT which can be used to sense larger displacements.

Angular displacement measurement methods fall into two groups – for very small angular displacements which are temporary, and for large displacements. Temporary small displacements are of a degree or less, and the shaft or other rotating object returns to its original position following the rotation. Such small displacements are sensed by variants of the methods used for small linear displacements, such as capacitive, strain gauge and piezo sensors. For larger displacements, inductive (particularly LVDT) methods are useful, and Fig. 2.25 shows a typical method. Potentiometric methods are also useful, but if the normal type of potentiometer construction is used, the angle of rotation is restricted to about 270 degrees. The rotary digital encoder can be used to provide a direct digital output, using a wheel of the form shown in Fig. 2.26 that has transparent and opaque sectors drawn in a pattern that

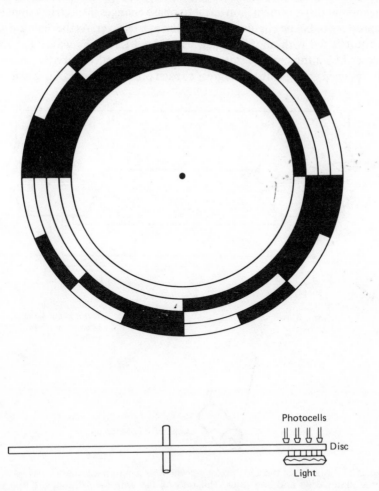

Fig. 2.26 A digital rotary optical encoder disk. In this example, only a sixteen position disk is shown, but practical examples would use eight tracks to obtain 256 positions.

provides as many channels of digital bits as will be needed for the required resolution.

The digital encoder can also be used to measure angular velocity and acceleration, and this makes it particularly useful for a very wide range of applications, assuming that suitable equipment is available to obtain the velocity and acceleration values from the angular position digital codes. This usually implies input to a computer that can run suitable software. If digital methods cannot be used, the changing output from another type of angular velocity measurement can be differentiated by means of an analogue computer stage using an operational amplifier.

One particularly useful sensor for angular displacement is the synchro, also known as the selsyn. This makes use of inductive principles, and a simple synchro system is illustrated diagramatically in Fig. 2.27. The aim is to sense an angular displacement, convert it to phase changes in electrical signals, and reproduce the same angular displacement at a receiver. As the diagram shows, a rotor is fed with an AC signal, usually a 1 KHz sine wave. The rotor is encased by a three-phase stator, with the coils equally spaced so as to make the induced voltages at 120 degrees to each other. The coils of this transmitter

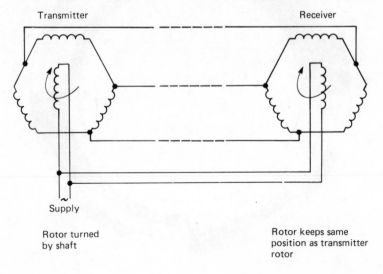

Transmitter   Receiver

Supply

Rotor turned   Rotor keeps same
by shaft   position as transmitter
                                      rotor

Appearance of each unit

Fig. 2.27 Principle of the synchro (or selsyn). This is a form of rotary LVDT which uses three stator coils and one rotor. The rotor is fed with an AC supply (often at high frequency), and the connection of the stator coils ensures that when one rotor is moved, the other (slave) units will respond with an identical rotor movement.

synchro are connected to the corresponding stator coils of a receiver synchro, whose rotor is fed with AC whose phase is locked to the phase of the AC used for the transmitter.

Any movement of the rotor of the transmitter synchro will cause a change in the amplitudes and phases of the voltages induced in the stator coils of the transmitter, and these same voltages and phases will exist across the receiver coils. The effect will be to cause a rotational force on the rotor of the receiver that will not reach zero until the rotor is at the same angular position relative to its stator coils as exists in the transmitter. The device can have very low friction, and because the transmission is electrical, and because each unit is fed from a mains supply, the sensitivity can be high and the amount of torque fairly large. The device is used for such diverse purposes as to transmit the angular position of a radar aerial, the direction of a wind vane, or the reading of a compass. The use of a synchro transmitter connected to a rotating radar aerial along with three-phase coils used for deflecting a cathode ray tube beam (with no rotor coil) was the main method of implementing the PPI (plan-position indicator) type of radar for many years.

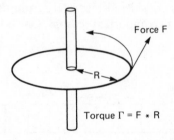

Fig. 2.28 Relationship of force, torque and radius for a revolving object.

Finally, the most difficult of the rotational quantities to measure is torque, the rotational equivalent of force. As Fig. 2.28 shows, torque can be obtained by using measurements of force or of angular acceleration. Of these, the force measurement, using a strain gauge mounted on a shaft so as to measure strain in the direction of the circumference of the shaft, is often the most practical arrangement for analogue measurements, since strain is proportional to stress caused by force. For a digital system, the use of an angular digital encoder and computer allows torque readings to be calculated.

Chapter Three
Light and Associated Radiation

Nature of light

Light is an electromagnetic radiation of the same kind as radio waves, but with a very much shorter wavelength and hence a much higher frequency. The place of light in the range of possible electromagnetic waves is shown in Fig. 3.1, along with an expanded view of the portion of the range which we class as infrared, visible and ultraviolet. This range of wavelengths is the subject of this chapter, and is considered separately from the longer wavelengths of electromagnetic for two important reasons. One is that this range of waves is

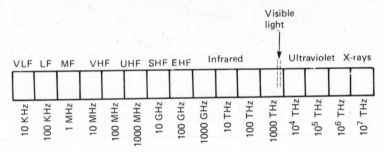

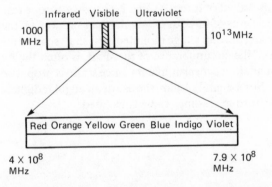

Fig. 3.1 The spectrum of electromagnetic waves ranging from 10 KHz upwards. Light is a very small part of this complete spectrum, and the waves that can be used for radio communication are at the lower frequency end.

not generated or detected by the conventional electronic methods that are used for waves in the millimetre to kilometre range. It would make no sense to talk of sensors and transducers for radio waves in this book, because no conversion is involved. The other point is that the very short wavelengths of this range cause effects that are not a problem when we work with radio waves. One such problem is coherence, mentioned earlier in conjunction with laser interferometers.

Like any other form of radiated wave, light can be polarised, and this is a topic that is of importance in some applications. Normal unpolarised light

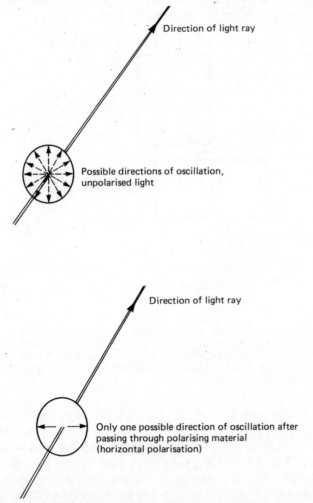

Direction of light ray

Possible directions of oscillation, unpolarised light

Direction of light ray

Only one possible direction of oscillation after passing through polarising material (horizontal polarisation)

Fig. 3.2 Light and polarisation. Light is an electromagnetic radiation which consists of an electric oscillation and a magnetic oscillation. The electric oscillation (a) can be directed anywhere at right angles to the motion of the light, but a polarised beam (b) has its direction of electric oscillation fixed. The magnetic oscillation is always at right angles to both the electric oscillation and the direction of the beam.

consists of waves whose direction of oscillation can be in any plane at right angles to the direction of motion (Fig. 3.2). When such light is passed through a polarising material, the light can become plane-polarised, meaning that the oscillation is in one plane only. Materials which polarise light in this way are generally crystals or other materials which contain lines of atoms at a critical spacing. If polarised light is beamed on another sheet of polarising material, the amount of light that passes through the material depends on the angle of the sheet, because two sheets of polarising material with their planes of polarisation at right angles will not pass any light (Fig. 3.3). Like radio waves, light can also be polarised in other ways (circular, elliptical polarisation), but plane polarisation of light is the most common.

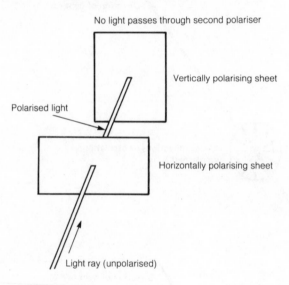

Fig. 3.3 Polarisers and their effect. Several natural crystals, and some synthetic materials, can plane-polarise light, confining the direction of the electrical oscillation to one direction. When two sheets of such materials are placed in the path of a light beam and arranged so that their directions of polarisation are at right angles, no light will pass through the second polariser.

Like the other waves in the electromagnetic range, light (considering the broad range of infrared to ultraviolet) has a velocity in space that is fixed at about 3×10^8 m/s, so that the usual relationship between wavelength and frequency (Fig. 3.4) holds good. When light travels through transparent materials, however, the velocity is reduced, and the factor by which the velocity is reduced is called the refractive index, always a number greater than unity. The velocity of light in any transparent material is therefore equal to the free-space velocity of 3×10^8 divided by the refractive index value for the material. For visible light, refractive index values for materials range from just above unity to about 2 – the highest values are for gemstones, notably

$$f \times \lambda = 3 \times 10^8 \qquad \text{(in free space)}$$

$$f = \frac{3 \times 10^8}{\lambda} \qquad f - \text{frequency in hertz}$$

$$\lambda = \frac{3 \times 10^8}{f} \qquad \lambda - \text{wavelength in metres}$$

Note: Speed of light and other waves of this family is lower in media such as glass.

Fig. 3.4 The relationship between wavelength, frequency and speed of waves. The figure of 3×10^8 m/s is for waves in free space (vacuum or air), and the speed can be significantly slower in other materials.

diamond. Refractive index values for infrared can be greater than this, and materials that are opaque to visible light may be quite transparent to infrared. This illustrates how critical the effect of the wavelength of the radiation can be. In general, the interaction between an electromagnetic radiation and a material can be expected to be critical when the wavelength of the radiation is of a value similar to the distance between atomic particles in the material.

Light radiation carries energy, and the amount of energy carried depends on the square of the amplitude of the wave. In addition, the unit energy depends on the frequency of the wave. This concept of unit (quantum) energy is seldom one that enters into consideration when longer wavelengths are being used, but it determines to a very considerable extent what can be done using light waves and particularly the sensing and transducing actions. The quantum nature of light will be explained in more detail when we consider the action of the vacuum photocell, for which the effect was first discovered and explained at the turn of the century. The explanation of photoelectric emission, incidentally, was the achievement for which Albert Einstein won his first Nobel prize.

Light flux

The sensitivity of photocells can be quoted in either of two ways – as the output at a given illumination, using illumination figures in units of lux, often 50 lux and 1000 lux, or as a figure of power falling on the cell per square centimetre of sensitive area, a quantity known as irradiance. The lux figures for illumination are those obtained by using photometers, and a figure of 50 lux corresponds to a 'normal' domestic lighting level good enough for reading a newspaper. The level of illumination required for close inspection work and the reading of fine print is 1000 lux; on this scale, direct sunlight registers at about 100 000 lux. The use of milliwatts/cm^2 looks more comprehensible to anyone brought up with electronics, but there is no simple direct conversion between power per square centimetre and lux unless other quantities such as spectral composition of light are maintained constant. For the range of wavelengths used in photocells, however, you will often see the approximate figure of $1 \text{ mW/cm}^2 = 200$ lux.

Another important point relating to the use of photocells is their peak sensitivity wavelength. For many types of sensors, this may be biased to either the red or the violet end of the visible spectrum, and some sensors will have their peak response for invisible radiation either in the infrared or the ultraviolet. A few devices, notably some silicon photodiodes, have their peak sensitivity for the same colour as the peak sensitivity of the human eye. This makes the devices more suitable for use in applications in which a process that once involved visual inspection is automated, but such replacements are not always successful. The reason is that though the peak sensitivity of the sensor may match that of the eye, the sensitivity at other colours may not follow the same pattern as that of the eye. In general, most sensors are more sensitive than the eye to colours at the extremes of the spectrum, and if a photosensor is to be used in applications such as colour matching then filters will have to be placed in the light path to make the response curve over the whole spectrum match that of the eye more evenly.

Photosensors

Photosensors are classed by the physical quantity that is affected by the light, and the main classes are photoresistors, photovoltaic materials, and photoemitters. Historically, photoemitters have been more important in unravelling the theory of the effect of light on materials, but photovoltaic materials, notably selenium, were in use for some considerable time before the use of photoemission became practicable. Since photoemission allows us to combine the description of a usable device with the quantum effect, we'll start this chapter by considering this type of sensor, which can also be used to a limited extent as a transducer. The photoemissive cell, in fact, was the dominant type of photosensor for many years, and played a vital part in cinema sound since it was the transducer for converting the film soundtrack into audio-electrical signals.

A simple type of photoemissive cell is illustrated in Fig. 3.5. This is a vacuum device, not least because the photoemissive material is one that oxidises instantly and violently in contact with air, and even in quite low pressures will oxidise sufficiently to prevent any photoemission. The photoemissive action is the release of electrons into the surrounding vacuum when the material is struck by light. This release of electrons will take place whether the electrons have anywhere to go or not, but unless there is a path provided, all the electrons will lose energy and return to the emitting surface, often termed the photocathode. By enclosing another metal – a nickel wire, for example – which is at a voltage more positive than the photocathode, the

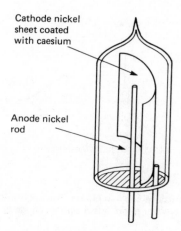

Cathode nickel
sheet coated
with caesium

Anode nickel
rod

Fig. 3.5 A typical photoemissive cell. The cell must be contained in an evacuated enclosure, and consists of a nickel rod anode with a nickel sheet cathode coated with a photoemissive material such as caesium.

cell can be used as part of a circuit in which current will flow when light strikes the photocathode. When light releases electrons, the current that flows is proportional to the amount of energy carried by the light beam.

This, however, is true only if the light is of a frequency that will release electrons. For any photoemitting material there will be a threshold frequency of light. Below this threshold frequency electrons will not be emitted no matter how intense the light happens to be, so that, in general, photoemitters do not respond to infrared, particularly the far infrared. The explanation of this threshold effect is due to Einstein, and makes use of an idea that was earlier put forward by Planck. Planck's theory was that energy existed in units, rather as materials exist in atoms, and he named the unit the quantum. The size of the quantum for a light beam is equal to the frequency of the light multiplied by a constant that we now call Planck's constant. We have already touched on this idea when considering laser interferometry, because the factor that prevents light from most sources from being coherent is that it is given out in these quantum-sized packets rather than as a truly continuous beam.

Einstein reasoned that the size of the quantum affected the separation of an electron from an atom in a photoemitter, and if the amount of energy carried in one quantum was less than the amount of energy needed to separate an electron, then no separation would take place. The total amount of energy carried by the beam was of no importance if the units of energy were insufficient to separate the electrons. The theory was confirmed by experiment, and this work also established Planck's quantum theory as one of the main supports of modern physics. The practical effect as far as we are concerned is that photoemissive cells have a limited range of response, and whereas it is easy to make cells that sense ultraviolet light (high frequency), it is much more difficult to prepare materials that will sense infrared. Most

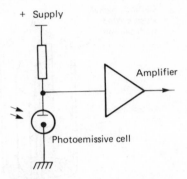

Fig. 3.6 The use of a photoemissive cell in a circuit. The supply voltage is in the range +25 V to +300 V, with a load resistor to provide a signal voltage to the amplifier.

photocells are noticeably much more sensitive to light in the blue/violet end of the spectrum than in the orange/red end. Mixed photocathode materials containing antimony along with the alkali metals caesium, potassium and sodium have been the most successful emitters in terms of providing an electrical output that is reasonably well maintained for the visible frequencies of light. This does not imply, however, that the output is by any means uniform.

The photoemissive cell is used in a circuit of the type shown in Fig. 3.6, with a voltage supply that is often in the region of 25 V to 100 V between anode and cathode. Some circuits make use of the current through the cell directly, but the most common circuit is as shown here, using a load resistor in series with the cell and amplifying the voltage signal. This is particularly useful when the incoming light is modulated in some way so that the electrical output will be an AC signal, such as that in use as a cinema soundtrack transducer. Where a DC output is needed, the use of a photoemitter is less simple, and a current amplifier is more useful than a load resistor and voltage amplifier. The current through a typical cell is of the order of a microamp, so that fairly large load resistors and a considerable amount of amplification will be needed. This creates difficulties with both frequency response and noise level when the photocell is used to convert modulated light signals into AC signals.

For measuring purposes, any photoemitter detector will have to be calibrated for each light colour for which it will be used, and if a reasonably high precision is needed, this calibration will be a long and tedious matter. It is important to realise that 'white' light is a mixture of all the frequencies of the visible spectrum, and can also contain a proportion of invisible ultraviolet. This means that imperceptible changes in the composition of 'white' light which have no effect on its total energy will nevertheless have very large effects on the output from a photoemissive cell. This is, in fact, something that affects most photosensitive devices, as keen photographers will know. Comparisons of light levels must therefore be made only when the composition of light is constant, and this is something that is very difficult to

achieve, particularly for natural lighting. For artificial lighting using
fluorescent tubes, constant light composition is much easier to achieve, but
filament lamps give a light that contains a large fraction of red light and for
which the light composition varies very sharply as the voltage is altered. A
filament lamp run below its rated voltage gives light that is predominantly
red; run above its rated voltage it can give light that is biased to the blue end
of the spectrum (with a greatly reduced life).

Photoemitters are still used where a fast response is needed, because many
competing devices are solid-state rather than high-vacuum, and the speed of
electrons in a solid is very much lower than the speeds that can be obtained
in a vacuum. Even for cinema soundtracks, however, the vacuum photocell
has now been replaced, and at the time of writing, vacuum devices are found
only in old equipment and in instruments intended for specialised use. For
these reasons, then, more detailed descriptions of vacuum photoemissive cells
will not be given here. However, the photomultiplier, is a method of obtaining
very much greater sensitivity from a photocell at very little cost in noise or
time delay. The principle is illustrated in Fig. 3.7, and makes use of secondary
emission from the same type of materials as are used for photocathodes,
notably caesium. Secondary emission occurs when a material is struck by
electrons and releases more electrons than strike the surface initially. The

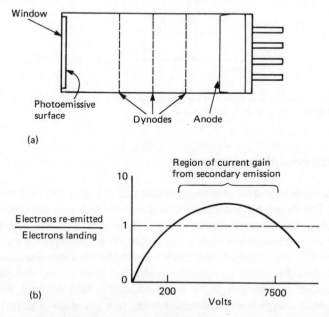

Fig. 3.7 (a) A photomultiplier using three dynodes. The electrons emitted from the
transparent photocathode are multiplied in the dynode stages to provide a much greater
output current than would be obtained from the original photocurrent. (b) The secondary
emission characteristic for a typical dynode material. As it happens, materials which are
good photoemitters are usually good secondary emitters as well.

effect is at its peak for electrons that have been accelerated by a voltage in the range of 30 V to 200 V, and for surfaces of caesium the multiplication factor can be large, 3 to 7. This means that an electron beam from a cathode can be directed to a secondary emitting surface and be re-emitted as a beam that contains a much larger number of electrons. In practical terms this means a beam at a higher current. By cascading stages, this multiplication effect can be very large, so that, for example, five stages that each give a multiplication of 5 will increase the current of a beam from a photoemitter by a factor of $5 \times 5 \times 5 \times 5 \times 5 = 3125$. This is unique among amplification methods in being virtually noiseless, so that the use of photomultipliers is a feature of detectors for very low light levels for specialised purposes. The secondary multiplying electrodes are known as dynodes, and each dynode must be run at a voltage level that is substantially greater than the one preceding it – Fig. 3.8 shows a typical DC supply arrangement.

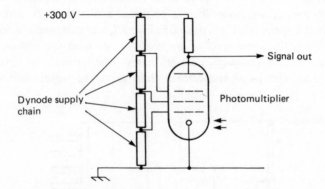

Fig. 3.8 The electrical arrangement for the photomultiplier, using a chain of resistors (1 M to 12 M values) to supply the dynodes.

Photoresistors/photoconductors

Many materials have a resistance value that will change when light strikes the material. The theory behind this effect is that these materials are semiconductors which in their normal state have few free electrons or holes. The effect of light is to separate electrons from holes and so allow both types of particles to move through the material and carry current. Since a definite amount of energy is needed in order to separate an electron from a hole, the size of the light quantum is important, but it is not difficult to find materials for which the amount of energy is small, corresponding to a quantum of infrared light. Since, in addition, the materials are comparatively easy to prepare in a practical form, the use of photoresistive or photoconductive materials is very common. The two names should really be synonymous, but some lists show devices under one name or the other. Because of the physical action, the effect is always that the material has high resistance in the absence of light, and the resistance drops when the material is illuminated. The effect of light on a photodiode is really the same type of action, and the reason for the use of a separate name is that the manufacturing process for these latter devices is the same as is used for other semiconductor devices.

The most common form of photoconductive cell is the cadmium sulphide cell, named after the material used as a photoconductor. This is often referred to as an LDR (light-dependent resistor). The cadmium sulphide is deposited as a thread pattern on an insulator, and since the length of this pattern affects the sensitivity, the shape is usually a zig-zag line (Fig. 3.9). The cell is then

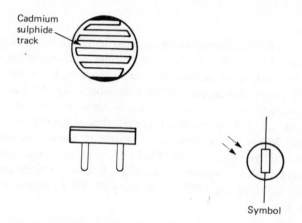

Fig. 3.9 The photoconductive cell, using a cadmium sulphide track, is by far the most common type of photosensor used industrially. It is mechanically and electrically rugged, and has a good record of reliability. The alternative name is LDR (light-dependent resistor).

encapsulated in a transparent resin or encased in glass to protect the cadmium sulphide from contamination by the atmosphere. The cell is very rugged and can withstand a considerable range of temperatures, either in storage or during operation. The voltage range can also be considerable, particularly when a long track length of cadmium sulphide has been used, and this type of cell is one of the few devices that can be used with an AC supply. Figure 3.10 shows typical specifications for a popular type of cell, the ORP12. The peak spectral response of 610 nm corresponds to a colour in the yellow-orange region, and the dark resistance of 10 M will fall to a value in the ohms to kilohms region upon illumination. The sensitivity is not quoted as a single figure because the change of resistance plotted against illumination is not linear. The maximum voltage rating of 100 V DC or AC peak allows a considerable latitude in power supplies, subject only to the 200 mW maximum dissipation.

The peak response at the yellow-orange region of the spectrum makes this LDR particularly useful for detecting the presence of a flame. The ORP12

Peak spectral response	610 nm
Cell resistance at 50 lux	2.4 kΩ
Cell resistance at 1000 lux	130 Ω
Dark resistance	10 MΩ
Max. voltage (DC or pack AC)	110 V
Max. dissipation at 25°C	200 mW
Typical resistance rise time	75 ms
Typical resistance fall time	350 ms

Fig. 3.10 The characteristics of the ORP12 type of photoconductive cell (LDR), courtesy of RS Components Ltd.

finds its main application in control units, in particular for vaporising oil-fired boilers (so that oil is cut off if the flame is extinguished) and in fire detecting equipment. In such applications, the main drawback of the LDR, a comparatively long response time, is not a drawback. The typical resistance fall time, meaning the time for the resistance to reach its final value when the cell is illuminated, is 350 ms; the resistance rise time when illumination is cut off is typically 75 ms. These times are reasonable when compared to the operating times of the relays that the ORP12 is so often used to operate, but they make any type of use with modulated light beams out of the question. A typical circuit utilising the ORP12 (courtesy of RS Components Ltd) is illustrated in Fig. 3.11.

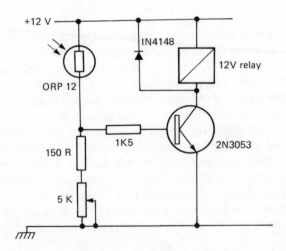

Fig. 3.11 A typical ORP12 industrial application circuit, courtesy of RS Components Ltd.

Photodiodes and phototransistors

A photodiode is a type of photoresistor in which the incident light falls on a semiconductor junction, and the separation of electrons and holes caused by the action of light will allow the junction to conduct even when it is reverse-biased. Photodiodes are constructed like any other diodes, using silicon, but without the opaque coating that is normally used on signal and rectifier diodes. In the absence of this opaque coating, the material is transparent enough to permit light to affect the junction conductivity and so alter the amount of reverse current that flows when the diode is reverse-biased. Because this is a diode reverse current, its amplitude is not large, and the sensitivity of photodiodes is quoted in terms of μA of current per mW/cm^2 of incident power. For the normal range of illuminations, this corresponds to currents of 1 nA to 1 mA at a reverse bias of –20 V. The lower figure is the dark current, the amount of current that flows with no perceptible illumination. As with all semiconductor devices, the dark current of a photodiode will increase considerably as the temperature is increased, doubling for each 10°C rise in temperature.

Peak response wavelength	750 nm
Sensitivity (typical)	0.7 μA/mV/cm^2
Dark current (–20 V supply)	1.4 nA
Temperature coefficient of dark current	$\times 2$ for 10°C use
Reverse breakdown voltage	–80 V at 10 μA
Temperature coefficient of change of signal current	0.35%/°C
Max. forward current	100 mA
Max. dissipation	200 mW at 25°C
Capacitance at –10 V bias	12 pF
Response time	250 ns

Fig. 3.12 Characteristics of a typical general-purpose silicon photodiode, courtesy of RS Components Ltd.

Figure 3.12 shows the characteristics of a typical silicon general-purpose photodiode from the RS Components catalogue. The peak spectral response is at 750 nm, which is in the near infrared, and the sensitivity is quoted as 0.7 μA/mW/cm^2. The typical dark current at a bias of –20 V is 1.4 nA, which means that the minimum detectable power input is of the order of 2 μW/cm^2. The current plotted against the illumination gives a reasonably linear graph, and the response time is around 250 ns, making the device suitable for modulated light beams that carry modulation into the video signal region. Figure 3.13 shows the suggested circuit for using this type of photodiode along with an operational amplifier for a voltage output. The feedback resistor R will determine the output voltage, which will be RI, where I is the

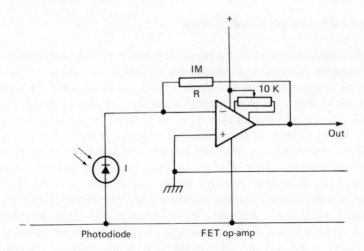

Fig. 3.13 A circuit making use of the silicon photodiode, courtesy of RS Components Ltd.

diode current. Note that when an operational amplifier is used in this way, the frequency response of the system is determined more by the operational amplifier and the stray capacitance across the feedback resistor than by the photodiode.

Phototransistors

A phototransistor is a form of transistor in which the base-emitter junction is not covered and can be affected by incident light. The base-emitter junction acts as a photodiode, and the current in this junction is then amplified by the normal transistor action so as to provide a much larger collector current, typically 1000 times greater than the output current of a photodiode. The penalty for this greatly increased sensitivity is a longer response time, measured in µs rather than in ns, so that the device is not suitable for detecting light beams that have been modulated with high-frequency signals. Phototransistors are used to a smaller extent nowadays because it is just as simple to make a chip containing a photodiode integrated with an operational amplifier, and better response times can usually be obtained in this way. The phototransistor is still found in conjunction with an infrared emitter used for such purposes as punched tape readers, end of tape detectors, object counters and limit switches.

Photovoltaic devices

The first form of photovoltaic device was the selenium cell, as used in early types of photographic exposure meters. The principle is that the voltage across the cell is proportional to the illumination, and since for the selenium cell the voltage was of an appreciable size (of the order of 1 V in bright illumination), an exposure meter using this type of cell needed no amplification and could use a meter of reasonably rugged construction. The use of selenium is by now of only historical interest – the properties of the metal were discovered by Wheatstone (of Wheatstone Bridge fame) when he was trying to construct high-value resistors to act as standards in a bridge circuit to measure the resistance of a trans-Atlantic telegraph cable. Modern photovoltaic devices are constructed from silicon, and the construction method is as for a photodiode. A silicon photovoltaic device is a silicon photodiode with a large area junction and used without bias. It is connected into a large load resistance, and the typical voltage output is of the order of 0.25 V for bright artificial illumination of 1000 lux. The main application of such cells is in camera exposure controls, because with suitable filtering the peak response and the response curve can be made very close to that for the eye. For the same reasons, the device can be used for the monitoring and control of light levels in critical manufacturing processes, and in instruments used for checking light levels. The photovoltaic cell can also be used as a photodiode, and if amplification is to be used, this mode is generally preferred to its use as a photovoltaic device.

Fibre-optic applications

The increasing use of optical fibres along with photosensors and optical transducers has led to the development of a range of devices which are specially intended for use with fibres. The detecting devices are photodiodes, so that the principles are the same as have already been discussed, but the physical form of each device must be suited to coupling to the fibre-optic cable. Fibre-optic links have become a practical proposition only since suitable terminations and connectors were developed, and it is futile to expect to be able to couple a general-purpose type of photodiode or emitter to a fibre cable. Several manufacturers offer complete fibre systems, composed of emitters, detectors, couplers and cables for either experimental work, assessment, or production use.

Light transducers

The conversion of energy between electrical and light forms has been of considerable importance for more than a century. The conversion of electrical energy into light has been and still is performed mainly by using the electrical energy to heat a coil of wire to a high temperature so that it gives out light. The light spectrum that is obtained in this way depends on the temperature of the object, and practical limitations usually result in the light having a spectrum that is biased to the red, so that the light from sources of this type, incandescent sources, is unsuitable for colour matching. This type of transducer is, however, the simplest to manufacture and use. It has also the lowest efficiency of conversion.

Early designs of light bulb used carbon filaments, which were fragile (some, astonishingly, have still survived), and the bulbs were evacuated to avoid oxidation of the carbon. The improvements made by Swan to the original Edison design included the use of the metal tungsten as a filament material, and the use of argon gas in the bulb. The inert nature of the gas avoided oxidation, and the pressure of the gas (though less than atmospheric pressure) avoided problems due to evaporation of the metal of the filament. Early tungsten-filament lamps failed in an hour or less because of the blackening of the glass caused by the tungsten vapour condensing. Though blackening is still the main cause of failure, the time has been greatly extended through the use of the inert gas in the bulb.

The development of gas-discharge lamps has resulted in a variety of light sources of a very different type, all very much more efficient than the incandescent lamp (usually at least five times the conversion percentage of the filament lamp). The original types of gas-discharge devices used gases such as neon at low pressure, so that high voltages in the range 1 to 15 kV were needed to operate the devices. These were and are the familiar neon signs, in which the gas is nowadays seldom neon but more usually a mixture of gases that have been chosen to achieve some particular colour. The colour of the light from discharge tubes is the colour of the predominant spectral lines of the gas or vapour that is used, and this leads to the light being in many cases almost monochromatic, of one single colour. Low-pressure discharge lamps are therefore unsuitable for general application to illumination and they remain used predominantly for advertising signs.

Vapours of liquids and solids have also been used in discharge lamps, and the most familiar of these are the mercury and sodium lamps that are used for street illumination. Used at low pressures, these vapours give strongly coloured light in which colour discrimination is almost impossible, but when higher pressures are used, the light spectrum broadens to give more acceptable results. The sodium high-pressure lamps are considerably more visible in fog, due to the orange-red predominance in the light, and are used extensively for motorway and main road illumination.

All varieties of discharge lamps require control gear, in the sense that they cannot simply be connected to a source of voltage. Figure 3.14 illustrates the normal current-voltage characteristic of a discharge lamp, and shows that no current passes until a critical voltage is reached across the terminals. From that point (the ignition voltage), the voltage across the gas drops as the current is increased, a negative-resistance characteristic. This would cause the lamp to burn out if there were no way of controlling the current, so that these lamps are normally operated with AC, using an inductor (choke) in series with the supply if the running voltage is less than supply voltage, or a transformer with

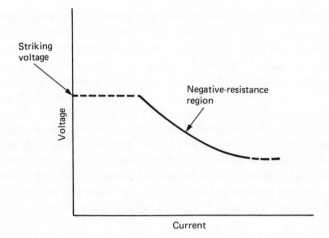

Fig. 3.14 A typical gas-discharge characteristic, showing the striking voltage and the negative-resistance region.

current-limiting inductance if the running voltage is higher than normal supply voltage. This additional equipment reduces the overall efficiency of the lamp, but is an essential part of the system.

For domestic use, the familiar fluorescent tube has been available for some considerable time now. This is not such a physically simple device as the older types of discharge tube, however, because it incorporates two energy conversions in series. The conversion from electrical to light energy is made by a discharge through mercury vapour at a pressure less than atmospheric, but this results in light that is mostly in the violet and ultraviolet range. The tube is coated on the inside with a phosphor material which acts as a frequency-changer. The violet/ultraviolet light from the mercury discharge causes energy changes in the atoms of the phosphor material, and as these changes are reversed the energy is given out again as light in the normal range of visible light.

This system is very flexible, because the colour and composition of the light that is given out depends on the phosphor coating rather than on the mercury discharge. The phosphors are metal silicates with controlled amounts of impurity, and very small changes in composition can cause marked differences in both conversion efficiency and spectrum – the materials are essentially the same as are used for cathode ray tube coatings. Because of this, fluorescent tubes can be bought in almost any colour, and most importantly in a 'sunlight' version whose spectral makeup is that of mid-day sunlight, ideal for colour matching. For domestic purposes, tubes labelled as 'warm' give a light that is more biased to red than the daylight type so that the light output is rather closer to that of the incandescent lamp. Even with the double conversion, the fluorescent is almost five times as efficient as an incandescent lamp.

The control gear of the fluorescent will be either of the switch-start or the quickstart variety. The switch-start type uses the conventional inductor (choke) in series with the tube, and a capacitor for power-factor correction. The starting voltage of the tube is higher than the supply voltage, and the mercury must be vaporised in order to achieve sufficient vapour pressure for starting. This makes starting difficult in cold conditions, which is why a fluorescent in a garage or a loft often fails to light in the winter. The starting circuit (Fig. 3.15) contains a thermal timeswitch that will pass current through the heating filaments when the tube is first turned on, and then break this circuit. At the instant when the switch breaks contacts, the back-EMF generated in the inductor causes the voltage across the tube momentarily to exceed the ignition voltage, so that the discharge starts and will then continue if the mercury vapour pressure is high enough. The current is then limited by the choke.

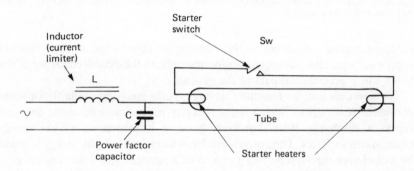

Fig. 3.15 The conventional switch starting circuit for a fluorescent tube, along with the inductor and power factor capacitor.

The alternative arrangement uses a transformer with separate windings for the heater filaments, and a high-impedance output at more than ignition voltage for the main discharge. At switch-on, the filaments are heated and ignition voltage is applied, so that the tube lights without the flickering or long delay that can be an annoying feature of switch starters. The high impedance of the transformer secondary then controls the tube current. A variation on this scheme uses an electronic converter to generate the heater and main voltages, so that the tube can be made with its control gear in one unit, and plugged directly into a domestic lamp socket. Tubes of this type are made by Wotan in Germany and are gaining in popularity for domestic use. Other tubes with incorporated control gear still use the switch-start principle and so exhibit the disadvantages of slow flickering starts. All domestic fluorescents of the compact type have a slow warm-up, so that the normal light output is not attained until the lamps have been in action for several minutes.

Solid-state transducers

Since the invention of the transistor, work on electronic devices that made use of current in solid semiconductors has resulted in the development of many useful devices, among which are the current to light transducers classed as LEDs (light emitting diodes). More recently, the place of the LED has been taken to some extent by a device that is, strictly speaking, neither a sensor nor a transducer but a light modulator, the LCD (liquid crystal display).

LEDs are diodes in which the voltage drop for conduction is comparatively large. When an electron meets a hole in such a junction, the two combine and release energy which can be radiated if the junction is transparent. The conventional germanium and silicon diodes release only a small amount of energy that corresponds to infrared, and which is absorbed in the non-transparent material (and also by the paint or other coatings on the diode), but compound semiconductors, of which the first was gallium arsenide, can give radiation in the visible range and are themselves transparent so as to allow the light to escape.

The majority of LEDs in use are of gallium phosphide or gallium arsenide phosphide construction and all are electrical diodes with a high forward voltage of about 2.0 V and a very small peak reverse voltage of about 3.0 V. In use, then, great care is needed to ensure correct polarity of connection so as to avoid burning out the LED. This has been made more difficult by the modern practice of identifying the cathode lead by making it shorter, because once the leads are cut to length for fitting into equipment, the identification is lost. The older form of identifying the cathode by a small flat area on the body of the diode was at least independent of the leads.

The intensity of illumination from the LED depends fairly linearly on the forward current, and this can range from 2 to 30 mA, depending on the physical size of the unit and the brightness required. All but the smallest types can sustain a power dissipation of the order of 100 mW, the main exceptions being the low-current LEDs that are used for battery-operated equipment. The current through the LED is governed by using a resistor in series, though when the LED is controlled by an IC, the resistor is usually incorporated into the IC and need not be wired separately.

The colour of the LED is determined by the material, and the two predominant colours are red and green. The use of red and green sources close to each other gives yellow light, in accordance with the rules on mixing of light colours, so that all three of these colours can be obtained from LED sources. The use of twin diodes in a package can allow switchable red, green and yellow lights to be obtained, and when LEDs are combined in a package with IC digital units, flashing LED action can be obtained.

Single LEDs are obtainable in a considerable variety of physical forms, of which the standard dot and bar types predominate. The bar type of LED can usually be obtained in intensity matched form, in which the intensity at a specified current rating is matched closely enough to allow the units to be stacked to be used as a column display. In a display of this type, a diode whose intensity is higher or lower than the others is particularly obvious, hence the need for intensity matching.

The other shapes of LED are not supplied as single units but as preformed patterns. Bar-graph modules of five to thirty segments can be bought as single units, but the most common preformed types are the alphanumeric displays. Of these, the seven-segment type of display, as illustrated in Fig. 3.16, is by far the most common. This uses seven bar-shaped segments to display numbers (and a few letters), and despite the name, most of the displays use eight segments because a decimal point is normally included and can be specified as on the right or on the left of the digit. The use of seven-segment displays requires a decoder-driver IC to convert from binary representation into the correct pattern of bars for each number.

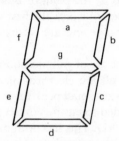

Fig. 3.16 The shape and labelling of the sectors of a seven-segment display. Many displays add an eighth segment, a decimal point to the right or left of the main sections.

The alternative to seven-segment display is dot matrix display. The dot matrix type of display can be used for numbers and letters, and if sufficient dots per unit are used, the display can be of any type of alphabet, not simply European alphabets. Dot patterns of 4×7 are used where only digits and capital (English) letters are needed, but 5×7 matrix patterns allow more versatility. Matrix displays are almost always sold as display units with controlling circuits built in, because there is seldom any requirement for a matrix display other than for the normal range of numerical and alphabetical characters (alphanumeric display).

Driving circuitry is also usually built in when multi-digit displays are used. Multi-digit displays are used on a multiplexed basis, meaning that display bars in identical positions on the different units are connected in parallel (Fig. 3.17). This would mean that a binary coded signal at the inputs would activate the same number on all of the displays, but for the multiplexing switches which isolate the cathodes of all the diodes except on one unit. On a four-digit display, for example, as might be used for a digital voltmeter, the digits are displayed one at a time, and the binary number data has to be altered at the input for each unit. This means that the rate of changing data at the inputs must be synchronised to the rate of switching from one display to another.

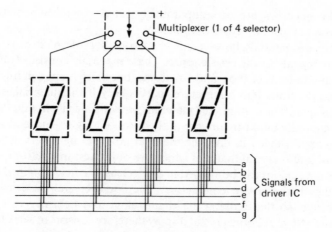

Fig. 3.17 Multiplexed displays. The signals from the driver IC are applied to the bars of all the displays in a set, and each digit is switched on in turn by a multiplexer, a form of IC switch that selects any one out of the four as illustrated here.

This usually calls for a display which is supplied without multiplexing circuitry, and an alternative is to use a display which incorporates a counter and multiplexing so that the pulses to be counted are input directly to the display which also has a reset and a blanking input.

One very considerable disadvantage of LEDs, used in any quantity, is power consumption. Even a small seven-segment display can use 15 to 20 mA per segment, so that the worst-case consumption for a four-digit display showing the number 8888 would be at least 420 mA. This order of current is reasonable enough for mains-powered equipment, but is out of the question for battery-operated circuits, unless secondary cells are being used (as on car displays, though these are usually turned off when the ignition switch is off). The display type that is much more likely to be used for battery-operated equipment, ranging from pocket calculators to portable computers, is the LCD (liquid crystal display).

Liquid crystals

Liquid crystals are not necessarily liquids, and definitely not crystals, but the name has stuck and is considerably easier to remember than any more precise title. The name of liquid crystal was given to certain types of organic materials (chemicals found in living organisms) in the early part of this century because of their remarkable chemical structure. A crystal is an arrangement of atoms in which the precise order governs the behaviour of the material, making it hard, of high melting point, and often polarising light. Each atom in a solid crystalline material affects each of its immediate neighbours very strongly, and the materials are of comparatively simple chemical structure, often composed of just two types of atoms.

There are materials, however, in which some sort of order exists, but which are not crystals in the classic sense. These materials consist of units which contain hundreds of thousands of atoms, but there is enough interaction to arrange the units into a structure, particularly if the material is a viscous (thick) liquid. The units in such materials, one of which is cholesterol, the fatty material in the bloodstream, are in the form of long chains, and when these chains are arranged in line, the material polarises light very strongly. The types of liquid crystal material which are of interest in electronics are those whose long chains can be aligned in an electrostatic field obtained by placing a voltage between two conductors. The liquid crystal materials are non-conductive, so that making the chains align in a field like this causes no current to flow. A liquid crystal display therefore consists of sets of electrodes with the liquid crystal material covering them, and with one transparent enclosing wall and a reflective back-plate (Fig. 3.18).

The liquid crystal display cell is used with a sheet of polarising material over the transparent wall. External light will pass through the polarising sheet, and if the liquid crystal material is not aligned, will pass also through this and be reflected from the back-plate. If a voltage is applied between two electrodes, however, the material between these electrodes will become polarising, and the already polarised light will not pass and cannot be reflected. This makes the affected area look dark, and the stronger the illumination of the cell the greater the contrast between this dark area and the brighter parts that have not been polarised. Though some displays are made with a built-in backlight for viewing in the dark, LCDs generally are intended for use in well-illuminated areas. The use of DC on the electrodes causes irreversible changes to the liquid crystal material, so that the usual practice is to invert the DC supply to low-frequency (30 to 60 Hz) AC and use this as the supply to the electrodes. The use of electronic conversion can ensure that no trace of DC is present, and since the power consumption is so low, relatively simple inverter circuits can be used.

Because the current requirement of the LCD is so low, of the order of 8 μA for a four-digit display operating at 32 Hz, there is no problem about using battery operation, even when a complex driving IC is incorporated with the LCD. It is quite rare to find LCDs offered solo – practically every manufacturer supplies only the packaged units with inverters and driving logic built in. Recently, the 'supertwist' type of LCD has become available, which offers much greater contrast along with faster operation, and this latter type is being used extensively for the screen displays of portable computers.

Other principles have been used, and are still used, for alphanumeric display. Filament displays, for example, can be obtained in seven-segment form, often with a better ratio of brightness to consumption than LEDs. Gas plasma displays, which were very common (in the form of the Ericsson Dekatron) prior to the emergence of LEDs, are still used for some portable

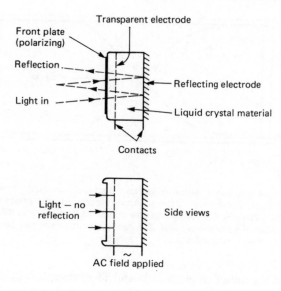

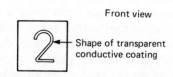

Fig. 3.18 The LCD principle. The shape that is to be displayed is deposited in the form of a transparent conductive coating on the front face, and the reflective rear face is the other electrode. When power is applied, the material between these electrodes polarises light, and the shape of the polarised zone accurately follows the shape of the electrode. Where the material is polarised, no light is reflected, causing a dark display.

computer screens because they allow viewing in poor illumination along with very low power consumption. The LED and LCD types, however, are by far the dominant types.

Light valves

A light valve is a device which is the electronic equivalent of a camera shutter. The LCD is one form of light valve, but the term is generally reserved for very high-speed devices that are capable of nanosecond timing. Light valves of this specialised type are used for ultra-high-speed photography, particularly for the photography of explosions and very fast chemical or nuclear reactions. The device is, like the LCD, a light modulator rather then a transducer, but some mention is due here.

The original type of light valve was a spin-off from TV camera techniques,

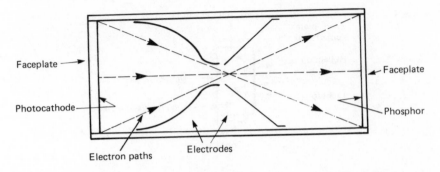

Fig. 3.19 A simple light valve, which makes use of an electron image. The electrons from the photocathode are accelerated to the phosphor screen to form a bright image. This can be turned off by changing the voltage on one or both electrodes, and the rate of turning the image on and off can be very fast. The same arrangement can be used as a light amplifier if the accelerating voltage is high enough.

and relied on the use of an electron beam. The principle is illustrated in Fig. 3.19, with a photocathode at one end of an evacuated tube and a screen at the other. With a suitable accelerating voltage between the photocathode and the screen, and possibly a magnetic field applied to maintain focus, the image on the photocathode can be obtained also on the screen. The electrodes near the photocathode can affect the electron beam, usually by deflecting it so that it does not reach the screen. This is most easily done if the beam is allowed to cross over or follow a bowed path (Fig. 3.20) on its way from the photocathode to the screen. High-speed electronic shutters of this type have been used to photograph events in nuclear explosions and are used extensively to analyse actions such as flame propagation in cylinderheads of car engines and in the ignition tubes of gas turbines.

The principle of photocathode and phosphor screen can also be used in image intensifiers. By using a large accelerating voltage between the photocathode and the screen, the light energy from the screen can greatly exceed the intensity reaching the photocathode. There can also be frequency conversion, because the photocathode can be an infrared-sensitive type and the screen a normal white one, allowing the capability to see images in apparent darkness. This principle was used in World War II as a gunner's night sight, and has been considerably refined since then in terms of sensitivity, size and power consumption and clarity of image.

Another well-established form of light switch is the Kerr cell. The Kerr cell uses an effect similar to the LCD – the effect of an electric field to rotate the plane of polarised light in a transparent material. Early types of Kerr cells used liquid suspensions, but later types have been able to use solid crystals. One application of modern Kerr cells has been to the modulation of laser light, allowing large-screen colour TV displays to be obtained from three laser beams. Another old-established effect, the Faraday effect, is used to deflect the light beams.

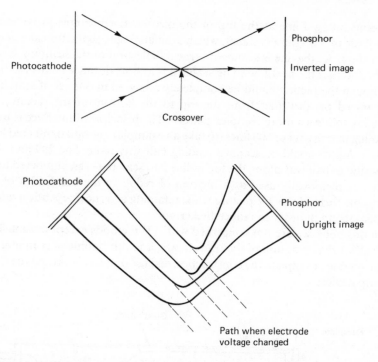

Fig. 3.20 Electron paths in light valves. In the crossover type, the image is inverted, and the action of the electrode is to turn off the beam. In the bent-beam type, image distortion is greater, but much less power is needed to restore a straight beam shape which means that no electrons reach the phosphor.

Image transducers

An image transducer is a device which will transform an optical image into electrical signals. The simplest image transducers are parallel types, in which a matrix of photodiodes supply signals to a set of wires. Until recently, this type of transducer would be used only for small-scale applications (such as optical character recognition) or coarse images, but modern developments in integration have made the principle interesting again. For the most part, however, image transducers are overwhelmingly of the serial output type in which the image is broken up into a set of repetitive signals.

The method of breaking up the image is scanning, a technique invented almost a century ago by Nipkov. Nowadays scanning means the deflection of an electron beam over a set of charges that are proportional to the illumination of parts of an image, but a serial-register action that gives a similar signal output is used in modern solid-state TV cameras and is also termed scanning. The basis of TV scanning is the analysis of the picture into lines and fields. Samples of the picture brightness are taken at a large number of points along a line at the top of the picture. Each sample provides the amplitude of a brightness signal, which is an analogue signal for each line of the picture. As the line is scanned, vertical deflection of the scanning is also applied so that the next line to be scanned is a short distance below the first.

Though the picture could be completely analysed in one set of scan lines that traced parallel lines from the top to the bottom of the picture, TV scanning follows a more complex pattern. The principle of interlace is used, meaning that in a set of 600 lines (to take an example), the odd numbered lines 1, 3, 5, 7, etc. would be scanned, ending half way along line 299 and then returning to half way along the top of line 2 to scan the even numbered lines. This was historically used as a method of reducing the bandwidth of the electronic form of the TV picture while retaining a vertical repetition rate of 50 per second in order to reduce flickering.

In this book we are not concerned with TV methods except inasmuch as they affect the operation of the transducers, and this outline of scanning has been intended to explain the construction and use of the vidicon and the CCD pickup devices.

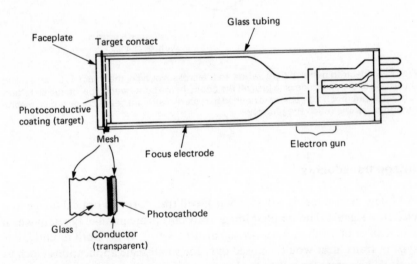

Fig. 3.21 Cross-section of a vidicon. The glass faceplate is coated with a transparent conducting material (tin oxide) and this in turn is covered with a layer of lead oxide photoconductor. The scanning beam brings the beam side of the photoconductor to cathode voltage, and it rises to target voltage by leakage through the photoconductor in the time between scan intervals. The voltage that is attained is proportional to the brightness of the illumination, and the discharge current when the beam scans across constitutes the output signal at the target electrode.

The vidicon principle is illustrated in Fig. 3.21. The sensitive material is a photoconductor, now traditionally lead oxide (antimony trisulphide was formerly used). The photoconductor is laid onto a transparent conductive coating which extends through the glass seal to metal contacts on the outside of the tube – it is from this connection that the video output signals are taken. The tube is evacuated, and the electron gun can produce a finely focused beam that can be deflected in the usual line and field pattern. Where the beam strikes the photoconductor, the connection will make the potential at the photoconductor equal to the potential at the cathode of the electron gun.

A light image is projected so as to focus on to the face-plate of the vidicon. This will have the effect of lowering the resistance of the photoconductor where the image is bright, allowing this portion to charge to the potential of the transparent conducting layer. The photoconductor behaves like a combination of capacitor and resistor (Fig. 3.22), so that between the times when the beam scans a point, the effect of the light is to build up an amount of charge that is proportional to the brightness of the light at that point. When the scanning beam strikes this spot, it will be discharged, and the capacitive discharge current will flow in the transparent layer, forming the video signal. This signal is at a comparatively high level, so that the vidicon is well suited to producing good signals in comparatively low illumination. Colour signals are obtained by using three vidicons, with filters used to make the images consist of the primary light colours of red, green and blue.

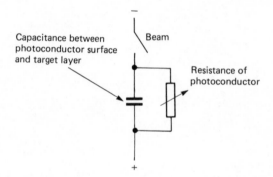

Fig. 3.22 The equivalent circuit of the photoconductor and target electrode arrangement.

The solid-state image transducer is considerably more complex, and is made in IC form. The principle is to use the light intensity to charge a capacitor and then to read the capacitor voltage by shifting it through a register. The digital type of shift register is unsuitable, and the device that is used is the CCD, charge-coupled device. This is a form of MOS device in which a clock pulse can pass a charge from one plate to another (Fig. 3.23), like an analogue form of shift-register. The CCD type of TV camera pickup device uses a large number of such devices which are arranged in lines. The lines are read out, giving a signal amplitude proportional to light brightness for each cell in the line, and a switching register ensures that the lines themselves are read in sequence. The picture quality does not approach that of the full-sized vidicon camera, but for very small cameras, or cameras that

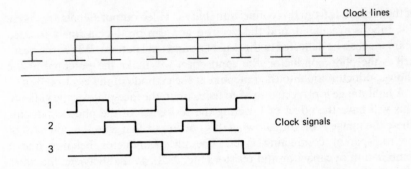

Fig. 3.23 CCD principles. The small contacts are charged by conduction through the underlying photoconductor. At scan time, a set of pulses on the clock lines will transfer the charges from contact to contact in the same direction, so that a serial output signal is obtained from the last contact, forming a video waveform for this line.

are used with low-resolution videorecorders (camcorders), the CCD type of device has the advantages of long life, high reliability, small size, and immunity to damage from bright lights.

Chapter Four
Temperature Sensors and Thermal Transducers

Heat and temperature – sensors

The physical quantity that we call heat is one of the many forms of energy, and its quantity is measured in the usual energy units of joules. The quantity of heat contained in an object cannot be measured, but we can measure changes of heat content that take place when there is a change of temperature or a change of physical state (solid to liquid, liquid to gas, one crystalline form to another). In this sense, then, temperature is a measure of the level of heat for a material whose physical state has remained unchanged. The relationship between temperature and energy is very similar to that between voltage level and electrical energy.

The temperature sensors that we use all depend on changes that take place in materials as their temperatures change. Transducers for electrical to thermal energy make use of the heating effect of a current through a conductor, but transducers for thermal to electrical energy are not so direct, and in accordance with the laws of thermodynamics will require a temperature difference to operate, taking heat in at a higher temperature and discharging some heat at a lower temperature.

The bimetallic strip

Thermal sensing is important for detection of effects as diverse as fire, overheating or the failure of a freezer. The simplest type of thermal sensor is the bimetallic type, whose principle is illustrated in Fig. 4.1. A strip is formed by riveting or welding two layers of metals, chosen so as to have very different values of expansivity. The expansivity (old name, expansion coefficient) is the

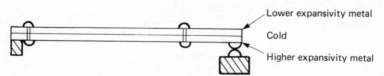

Lower expansivity metal

Cold

Higher expansivity metal

Fig. 4.1 The bimetallic strip consists of two metal strips welded or riveted together. The strip can be extended into a spring shape for greater sensitivity, or can consist of two welded discs which will buckle when heated.

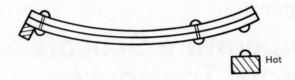

Hot

Fig. 4.2 If one metal has a higher expansivity than the other, the strip will bend when heated, with the metal of higher expansivity on the outer side of the sector of the circle.

fractional change of length per degree change of temperature and for all metals is positive, meaning that the strip expands as the temperature increases. Because the metals do not expand by the same amount, however, the strip will bend as the temperature changes, as indicated in Fig. 4.2.

This bending action can be sensed by a displacement transducer of any of the types discussed in Chapter 2, but is more often used to operate switch contacts, usually with the strip itself carrying one contact. The conventional type of bimetallic strip element is still to be found in some thermostats, though the strip is very often arranged into a spiral. This allows for much greater sensitivity, since the sensitivity depends on the length of the strip. The amount of deflection can be fairly precisely proportional to temperature change if the temperature range is small.

Thermostats of this type, however, have an undesirably large hysteresis, so that a thermostat set for a nominal 20°C might open at 22°C and close again at 18°C. This leads to undesirable temperature swings and so to occupants of a room ignoring the thermostat and turning radiators directly on or off, or using the thermostat simply as an on/off switch. The hysteresis of the simple bimetal thermostat can be reduced by the use of an 'accelerator', consisting of a resistor placed close to the element. The principle is that when the thermostat contacts close to switch on heating in a room, current is passed through the accelerator resistor (Fig. 4.3) so that the rate of heating within the thermostat is faster than outside. This leads to the thermostat points opening before the same temperature is achieved in the room outside. The current through the accelerator resistor then switches off, and the thermostat will then cool more rapidly than the room so that the switch on is more rapid than would otherwise be the case. The use of an accelerator, however, can lead to

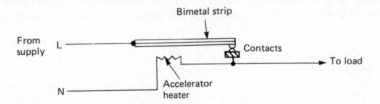

Bimetal strip

From supply

L

Contacts

To load

N

Accelerator heater

Fig. 4.3 Using an accelerator with a bimetal thermostat. The accelerator ensures that the rate of rise of temperature at the thermostat is greater than that of the surroundings, so overcoming the hysteresis of the thermostat to some extent.

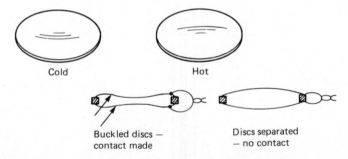

Cold   Hot

Buckled discs –
contact made

Discs separated
– no contact

Fig. 4.4 Bimetallic discs – these are used extensively as sensors for overheating
components such as transformer windings and electric motors.

the desired working temperature being achieved very slowly or not at all in
cold weather, and much too rapidly in hot weather. This has lead to the use
of more sensitive devices for thermostat use, based on thermistors (see later in
this chapter).

The bimetallic strip exists in several physical forms, and one particularly
useful form is the disc (Fig. 4.4). For a change of temperature, a bimetallic
disc will abruptly buckle, giving a snap-over action that requires no form of
assistance. This is the basis of the small thermal switches that are used for
overheating protection in electronic equipment that incorporates heat sinks,
small motors, transformers or other components that are likely to overheat
and have a metallic surface to which the thermal switch can be bolted. These
thermal switches can be bought as normally open or normally closed types,
depending on whether they are to be used to detect rising or falling
temperatures, with preset nominal temperatures that have temperature
hysteresis of the order of 3° to 5° on each side of the set temperature since no
accelerator is used. For more precise control, units that use long bimetal strips
can be obtained with smaller hysteresis and variable setting temperature. All
types of long-element bimetal strip thermostats should be recalibrated at
intervals, since the strip is subject to gradual changes (creep) that affect the
thermostat setting.

Liquid and gas expansion

Another principle used in temperature sensing is liquid expansion in
conjunction with a pressure switch, making use of the principles of the
familiar mercury thermometer. The simplest sensor of this type is an
adaptation of a mercury thermometer with two wire electrodes inset into the
capillary (Fig. 4.5). Since mercury is a conducting metal, a circuit will be made
through the electrodes when the mercury level reaches the electrode whose
position corresponds to the higher temperature. This allows a predetermined
temperature to be sensed, but for a switching action only and with no way of
altering the temperature at which switching takes place other than by
replacing the sensor with another one.

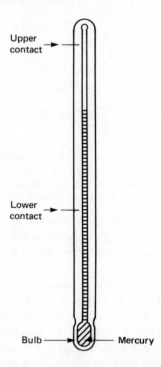

Upper
contact

Lower
contact

Bulb   Mercury

Fig. 4.5 A temperature switch developed from an ordinary mercury thermometer and using wire electrodes embedded in the glass tubing.

Though the mercury level can be used to change the frequency of an oscillating circuit and thus to provide a proportional sensing of temperature, this type of action is seldom used. The sensors that are used for temperature measurement as distinct from switching are mainly of the electronic type, including thermocouples and thermistors, and devices that make use of mechanical expansion are more likely to be used in switching circuits. The most common type is a development of the conventional bulb thermometer and has a sensing element (Fig. 4.6) consisting of a capsule filled with liquid that is connected by a narrow-bore tube to the pressure switch. The liquid need not be mercury, and is nowadays more likely to be a form of synthetic oil.

Since the capsule can be remote, and involves no electrical connections, this is often very useful for hazardous environments, and the liquid can be chosen accordingly. The length of connecting tubing must be such that the volume of liquid contained in the tubing is only a small fraction of the total volume, since the temperature of the liquid in the tubing will also affect the pressure. The use of air or an inert gas in place of a liquid makes the device very much more sensitive, but the pressure switch needs to be able to respond to much lower pressures than are exerted by an expanding liquid.

One disadvantage of the system generally is that the sensing capsule needs

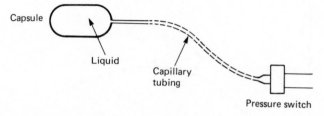

Fig. 4.6 A liquid bulb and pressure sensor method for temperature sensing. The volume of liquid in the capillary tubing should be negligible compared to the volume of liquid in the bulb.

to contain a reasonable volume of liquid, and so cannot be small. In addition, since this volume of material has to be heated and cooled in order to follow temperature fluctuations, time is needed for the change, so that capsules cannot follow rapidly changing temperatures. The pressure sensor need not be a switching device, and the use of a diaphragm coupled to a potentiometer, LVDT or piezoelectric transducer can make the liquid/bulb type of temperature sensor into a fairly precise instrument, though this type of use has few applications.

Electronic/electrical sensors

The thermocouple is frequently used as the sensing element in a thermal sensor or switch. The principle is that two dissimilar metals always have a contact potential between them, and this contact potential changes as the temperature changes. The contact potential is not measurable for a single connection (or *junction*), but when two junctions are in a circuit with the junctions at different temperatures then a voltage of a few millivolts can be detected (Fig. 4.7). This voltage will be zero if the junctions are at the same temperature, and will increase as the temperature of one junction relative to

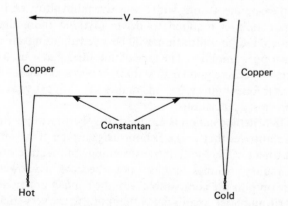

Fig. 4.7 The construction of a thermocouple, in this example using copper and constantan.

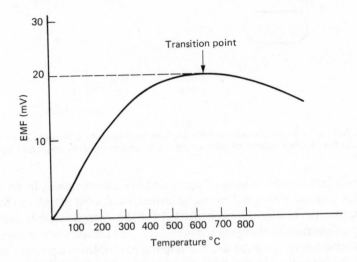

Fig. 4.8 A thermocouple characteristic, showing the typical curvature and the transition point at which the characteristic reverses. A few combinations of metals (like copper/silver) have no transition, but have a very low output.

the other is changed until a peak is reached. The shape of the typical characteristic is shown in Fig. 4.8, from which you can see that the thermocouple is useful only over a limited range of temperature due to the non-linear shape of the characteristic and the reversal that takes place at temperatures higher than the turn-over point.

The output from a thermocouple is small, of the order of millivolts for a 10C° temperature difference, and Fig. 4.9 shows typical sensitivity and useful range for a variety of the common types. Of these, the copper/constantan type is used mainly for the lower range of temperatures and the platinum/rhodium type for the higher temperatures. Because of the small voltage output, amplification is usually needed unless the thermocouple is used for temperature measurement along with a sensitive millivoltmeter. If the output of the thermocouple is required to drive anything more than a meter movement, then DC amplification will be needed, using an operational amplifier or chopper amplifier. The type of amplifier that is used needs to be carefully selected, because good drift stability is necessary unless the device is recalibrated at frequent intervals. This makes the chopper type of amplifier preferable for most applications.

If an on/off switching action is required, the thermocouple must be used along with a controller that uses a Schmitt trigger type of circuit which also permits adjustment of bias so that the switching temperature can be preset. The usual circuitry includes amplification, because the lower ranges of thermocouple outputs are comparable with the contact potentials (the same type of effect) in amplifier circuits, and attempting to use very small inputs for switching invariably leads to problems of hysteresis and excessive sensitivity.

One particular advantage of thermocouples is that the sensing elements

(a)

Metal	EMF (mV)	
Constantan	-3.3	
Nickel	-1.5	
Aluminium	0.4	
Manganin	0.65	
Silver	0.7	All figures for 100°C temperature
Copper	0.75	difference using platinum as the
Tungsten	0.8	other metal.
Molybdenum	1.2	
Iron	1.88	
Silicon	45.0	

(b)

Temp. °C	Copper/ constantan	Iron/ constantan	Platinum Plat/rhodium	
-20	-0.75	-1.03		EMF in mV,
-10	-0.38	-0.52		cold end at
0	0.00	0.0	0.0	0°C.
10	0.39	0.52	0.05	Only the
20	0.79	1.05	0.11	useful
30	1.19	1.58	0.17	range is
40	1.61	2.12	0.23	shown.
50	2.04	2.66	0.30	
60	2.47	3.20	0.36	
70	2.91	3.75	0.43	
80	3.36	4.30	0.50	
90	3.81	4.85	0.57	
100	4.28	5.40	0.64	
200	9.29	10.99	1.46	
300	14.86	16.57	2.39	
400	20.87	22.08	3.40	
500		27.59	4.46	
600		33.28	5.57	
700		39.30	6.74	
800		45.71	7.95	
900		52.28	9.21	
1000		58.23	10.51	
1200			13.22	
1500			17.46	

Fig. 4.9 The thermoelectric behaviour of metals. (a) The EMF at 100° C difference for platinum/metal thermocouples. (b) More detailed characteristics for three important types, showing the useful range of temperature differences and EMF values.

themselves are very small, allowing thermocouples to be inserted into very small spaces and to respond to rapidly changing temperatures. The electrical nature of the process means that the circuitry for reading the thermocouple output can be remote from the sensor itself. Note that thermocouple effects will be encountered wherever one metallic conductor meets another, so that temperature differences along circuit boards can also give rise to voltages which are comparable with the output from thermocouples. The form of construction of amplifiers for thermocouples is therefore important, and some form of zero-setting is needed.

Metal-resistance sensors

All metallic conductors exhibit a change of resistivity when their temperature changes, and this change in the resistivity causes a change of resistance – Fig. 4.10 summarises the relationships. The change of resistance is a more linear quantity over a large range of temperature than the output from a thermocouple, and though the characteristic shows deviation from a straight line at high temperatures, there is at least no reversal as is found in the thermocouple characteristic. The deviation is caused by the effect of the square and cube law components of the equation, and these effects are important only at high temperatures. For most metals, the first coefficient of resistance change (alpha) is close in value to the 1/273 (0.003 66) figure for the expansivity of gases. A few metal alloys have a very low value of temperature coefficient, notably constantan with a value that is only about 1% of the average for pure metals, and manganin with an even lower value. Both of these materials are alloys of copper, nickel and manganese.

For comparatively small temperature ranges, up to 400°C or so, the resistance change of nickel or of nickel alloys can be used, and for higher temperature ranges platinum and its alloys are more suitable because of their much greater resistance to oxidation. For measurement purposes, the resistance sensor can be connected to a measuring bridge, along with a set of dummy leads whose temperature is also changed (Fig. 4.11). A platinum resistance in this form can be used as a standard of temperature measurement. The National Physical Laboratory standard thermometer is a gas-expansion type, but this requires long and elaborate setting up, so that platinum-resistance thermometers calibrated from the gas-thermometer standard are used extensively as secondary standards (often called, confusingly, sub-standards). The size of the sensing element and its heat capacity make the response slow compared to some of the purely electronic devices, such as the thermocouple.

When a switching action is required, the output from a bridge circuit connected to the platinum resistance thermometer can be used to operate a trigger type of circuit. This is seldom done, because the advantage of the resistance type of thermometer is its comparatively linear response, and switching can be carried out by a variety of less expensive devices.

For a wire with cross-sectional area A, length s, resistivity ρ

the resistance R = $\dfrac{\rho s}{A}$

For a temperature change of θ °C, the following changes occur:

length increases by s.α.θ α = coefficient of linear expansion (or linear expansivity)

area increases by 2A.α.θ

resistivity increases by ρ.α'.θ α' = temperature coefficient of resistivity

For most metals α' is about 4×10^{-3} per degree
α is about 2×10^{-5} per degree

so that the changes in dimensions can be ignored.
We can therefore use α' as the temperature coefficient of resistance.

$R_\theta = R_0 (1 + \alpha'\theta)$ R_θ = resistance at temperature θ
R_0 = resistance at 0°C
α' = temperature coefficient of resistance
θ = temperature difference.

Metal	α'
Aluminium	4.2×10^{-3}
Copper	4.3×10^{-3}
Iron	6.5×10^{-3}
Nickel	6.5×10^{-3}
Platinum	3.9×10^{-3}
Silver	3.9×10^{-3}

Fig. 4.10 Resistance, resistivity and the change of resistance with temperature.

Thermistors

For the lower ranges of temperature of the order of 0°C to 150°C, electronic sensing methods based on semiconductors are used. The thermistor is typical of this type of device, and its change of resistance per unit change of temperature can be so abrupt that circuit devices such as bridges and Schmitt triggers are often not necessary. In a very few applications, the thermistor can be used directly, but it is usually undesirable to have the controlled current passing through the thermistor, and more usually the thermistor is part of a transistor switch or an operational amplifier circuit (Fig. 4.12). The output of such a circuit is not particularly linear, but the sensitivity can be high and the

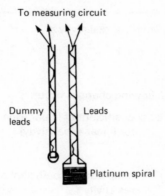

(a)

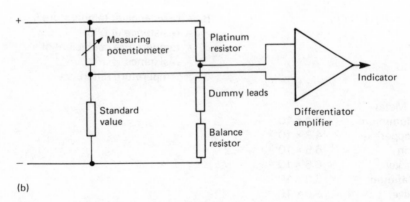

(b)

Fig. 4.11 The arrangement of a platinum resistance thermometer, using dummy leads to compensate for the change of resistance of the leads to the measuring element: (a) physical arrangement, (b) electrical circuit.

response can be rapid. One particular advantage is that the sensing element can be very small.

The thermistor itself is not a semiconductor of the germanium or silicon type but a mixture of semi-metal oxides, the type of materials sometimes described as 'rare earths' because they are found only in small quantities and in very few locations. The typical thermistor characteristic is shown in Fig. 4.13, and consists of a negative change of resistance with increasing temperature. The shape of the characteristic is exponential rather than linear, and the useful temperature range is comparatively small.

The connection of a thermistor to a switching circuit has the advantage, as compared to bimetallic strip devices, that it can be arranged so as to have zero

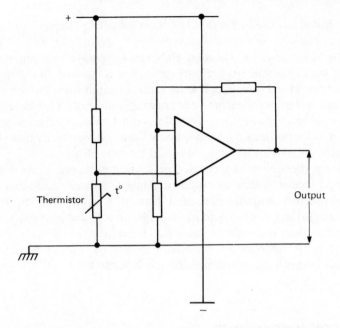

Fig. 4.12 A thermistor temperature sensing circuit that makes use of an operational amplifier. The sensitivity of the detector can be altered by changing the feedback ratio.

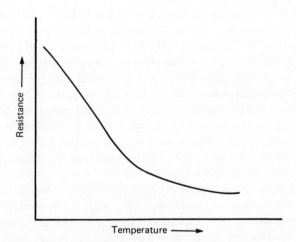

Fig. 4.13 A typical resistance/temperature characteristic for a thermistor. The resistance decreases as the temperature rises, and the shape of the characteristic is never linear.

hysteresis if this is a useful feature. For most switching purposes, however, hysteresis is desirable in order to prevent rapid switching on and off as air currents strike the detector. The more elaborate temperature sensing systems that make use of thermistors are microprocessor controlled, and a form of time hysteresis is used. The temperature sensing output from the thermistor is monitored at short intervals, and a change registered only if the direction of temperature change is consistent. This allows much more rapid response to a temperature change than the conventional form of hysteresis, though the sensing intervals have to be adjusted to suit the type of use.

Transistors themselves can be used as sensing elements, because many transistor parameters follow an exponential negative temperature characteristic. The inherent amplification of a transistor makes the use of the temperature sensitivity of base-emitter voltage attractive as a sensing system, because the output can be taken in amplified form from the collector. The use of transistors as temperature sensors, however, is confined mainly to temperature compensation of transistor and IC circuits.

Radiant heat energy sensing

Radiant energy, which can be light, heat, radio waves or in some instances ionising radiation, may need to be sensed, and a very large range of the electromagnetic spectrum is sensed by its temperature effect. The sensors for light have mainly been dealt with in the previous chapter, but there is one device, the bolometer, which has been held over to this chapter because its action is essentially thermal.

The bolometer principle is illustrated in Fig. 4.14. A blackened material will absorb radiation well and so its temperature will be raised when radiated energy falls on it. The change of temperature is then detected in the ways that have been described earlier in this chapter. The classic types of bolometers used in the nineteenth century were metal, and the effect of the temperature rise due to radiation was detected in sensitive measuring bridge circuits. Because the change of temperature due to radiated energy can be small, and the change of resistance correspondingly smaller, bolometers are usually connected into bridge circuits so that the output from an unradiated bolometer can be compared with the output from the bolometer which is deliberately exposed to radiation.

Modern bolometers can make use of semiconductor sensors, blackened for maximum absorption of radiation. The very much greater change of resistance of a thermistor for a small change of temperature makes this type of material ideal for bolometer use and permits much more sensitive detection than was possible with the older types. The non-linear nature of the thermistor is less important in this type of application, because the changes of temperature are usually small.

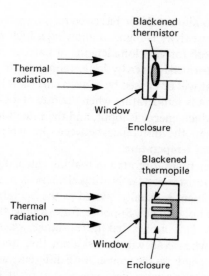

Fig. 4.14 The bolometer consists of a temperature sensor whose surface is blackened and enclosed in a container (preferably evacuated). Thermal radiation passes through the window, heating the sensor. The sensor can be a thermistor or a thermopile, an array of thermocouples connected in series.

Pyroelectric materials

More recently, the phenomenon of pyroelectricity has been used as a method of detecting changes of temperature, and in particular for the changes that are caused by radiant heating. A pyroelectric material is one which will generate an EMF in the bulk of the material (that is, not in junctions with other materials) in response to heating, and the modern pyroelectric sensors are all based on plastics of the polyethylene type. When such plastics are prepared in the same way as is used for electrets, by slowly solidifying in a magnetic field, they become highly pyroelectric, and if they are used as the dielectric of a capacitor, the EMF can be measured between the films of metal on each side of the plastic.

Such sheets can be blackened and used as very sensitive detectors of radiant heat, sufficient to be used in fire detectors. They are also vibration sensitive and can be used as a low-cost form of microphone (see Chapter 5), so that security devices can make more than one use of this principle.

Thermal transducers

The transducer for electrical to thermal conversion is the heating element, and for most purposes this is composed of a nickel alloy wire such as Nichrome.

This alloy of nickel, chromium and iron has good resistance to oxidation even when red hot, and its resistivity is high, so allowing a high resistance to be achieved without the need for very long lengths of narrow-gauge wire. The amount of energy transformed is given by Joule's equation (Fig. 4.15), but the level of temperature that will be caused by a given current is less predictable. A material reaches a steady temperature when the rate of loss of heat energy is equal to the rate at which energy is input, and the rate of loss depends on many factors other than the difference between the temperature of the material and the ambient temperature.

One point that is sometimes forgotten in making calculations on thermal conversion is that the resistance of a heating element is not constant. The current in a heating element is usually quoted as the operating current, the value of current for the hot element. Suppose, for example, that a heater has to operate at 240 V, 2 A. This implies that the resistance, *when hot*, will be 120 ohms. The resistance when cold will be less than this, and the difference between these values depends on the temperature difference and the material. If the heater of our example uses nichrome and runs at 400°C then the resistance at room temperature (taken as 25°C) is 113 ohms, because the temperature coefficient of nichrome is small. If the heating element had been constructed of pure nickel then the resistance at room temperature would be

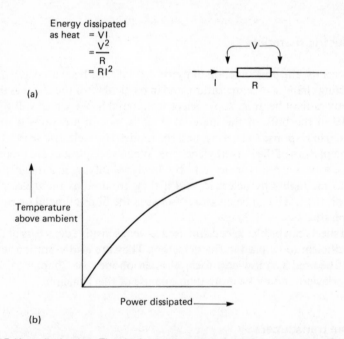

Energy dissipated
as heat = VI

$$= \frac{V^2}{R}$$

$$= RI^2$$

(a)

Temperature above ambient

Power dissipated ⟶

(b)

Fig. 4.15 Heat dissipation. The Joule equations (a) show the conversion of electrical energy into heat energy, but there is no straightforward way of finding what the final temperature of the hot object will be. A typical temperature/power curve for a resistor is shown in (b).

Resistance/temperature formula is:

$R_\theta = R_0 (1 + \alpha'\theta)$

R_θ = resistance at temperature θ
R_0 = resistance at 0°C
α' = temperature coefficient of resistance
Take α' as $7.7 \times 10^{-4} R^{-1}$

1. $R_{400} = R_0 (1 + 7.7 \times 10^{-4} \times 400)$
2. $R_{25} = R_0 (1 + 7.7 \times 10^{-4} \times 25)$

So that $R_{400} = R_0 (1.068)$
$\quad\quad\quad R_{25} = R_0 (1.00425)$

and $\dfrac{R_{400}}{R_{25}} = \dfrac{1.068}{1.00425} = 1.06$

If $R_{400} = 120 \, \Omega$ then $R_{25} = \dfrac{120}{1.06} = 113.2 \, \Omega$

Fig. 4.16 How resistance changes caused by temperature changes are calculated.

about 47 ohms, a very considerable change. Figure 4.16 shows the basis of these calculations. The changes in resistance are particularly marked for incandescent lamps, and some care should always be taken if you need to work with the resistance values of such lamps, since the cold value will be only a small fraction of the hot value.

Any conducting material will act as a transducer of electrical energy to thermal energy, so that the use of materials which have a negative value of temperature coefficient needs special care. Passing current through such a material from a fixed voltage supply can cause heating which in turn lowers the resistance and increases the current. This 'positive feedback' can cause fusing or other breakdown unless some part of the circuit limits the amount of current that can flow. The most notable examples of this effect are semiconductors and that curious mixture, glass.

Thermal to electrical transducers

The conversion of electrical energy to thermal energy proceeds with almost 100% efficiency, but no conversion from thermal energy to any other form ever approaches much more than 50% efficiency. The reasons for this are summed up in the laws of thermodynamics, and are founded on the principle that we do not know how much heat an object contains, and we cannot remove all of it. Any change from thermal energy to another form must involve heat taken in by a converter at a high temperature and a lesser amount of heat given out at a lower temperature. The efficiency cannot then be greater than the fraction given by:

$$\frac{\text{temperature difference}}{\text{high temperature}}$$

with the temperatures measured in the Kelvin scale whose zero point is equivalent to -273.16°C. This equation implies that 100% conversion is possible only if the converter exhausts its heat at 0 K, which is not a practical proposition. In addition, this equation assumes that every other part of the conversion process operates at 100% efficiency.

The conversion from thermal energy to electrical is, on the large scale, carried out by way of steam generation, with the steam operating turbines that are coupled to alternators. The source of thermal energy can be nuclear. The use of the more direct gas-turbine and alternator method is expensive and is used only for topping up the supply from conventional coal and nuclear power stations. One of the benefits of the inefficiency of the whole process can be the availability of large quantities of water at a useful domestic temperature of 40°C to 60°C, and in some countries electricity generating stations also sell their waste heat in a type of scheme called CHP, combined heat and power. In the UK, the emphasis has always been on large-scale generation and grid distribution, and CHP has never been a practical proposition in such circumstances. The pioneer dream of a small nuclear station supplying electricity and heating for a self-contained community has never been realised, but it will have to become an option as the effects of burning half of the contents of the planet become more noticeable.

By comparison with the 40% efficiency that can be obtained from a well-designed and very large coal-fired power station, the efficiency of the generation of electricity from any other thermal transducers is extremely low. The most practical system in the past has used thermocouples stacked into large blocks that are heated at one end and cooled at the other, with the individual thermocouples connected in series so as to obtain a useful voltage. Several commercial units have been available in the past, and at one time the gas-fired radio receiver was, if not common, more than a museum piece. The Milnes converter was one of the most successful of these units, and was manufactured in a factory in Tayport, Fife in the late forties, along with loudspeakers to the Milnes design. The efficiency of these thermocouple units, however, rarely exceeded 5%.

Other forms of thermal to electrical converters have achieved even lower efficiencies, and even with intensive development work the efficiency figure of 10% is more of a dream than a reality. One method that was pursued for conversion of solar energy used the principle of thermal emission from a cathode heated by focused sunlight, with a large number of units connected in series in order to attain a reasonable voltage. This, however, proved to be as unreliable as any other solar source, and the ultimate comment on solar heating is provided by the advertisements in Australian papers offering to replace solar heating units by oil-fired boilers. If solar heating is uneconomic in Australia, its chances in the rest of the world are rather poor.

Chapter Five
Sound, Infrasound and Ultrasound

Principles

Sound and vibration are connected in the sense that any sound is associated with a mechanical vibration at some stage. Many sounds are caused by the vibration of solids or gases, and the effect of a sound on the hearer is to vibrate the eardrum. The sound wave is the waveform caused by a vibration and which in turn causes an identical vibration to be set up in any material affected by the sound wave. The mechanical vibration need not necessarily cause any sound wave, because a sound wave needs a medium that can be vibrated, so that there is no transmission of sound through a vacuum. Perhaps some day this will penetrate the minds of the makers of 'space films', who always consider sound effects so important.

When sound is transmitted, the wave parameters are velocity (speed), wavelength and frequency. The frequency and the waveshape are determined by the frequency and waveshape of the vibration that causes the sound wave, but velocity and wavelength are dependent on the medium that carries the sound wave. The relationship of velocity, wavelength and frequency is illustrated in Fig. 5.1. The velocity of sound in a given material depends on the density and the elastic constants of the material, and the velocity is highest in dense solids, lowest in gases at high pressure. Figure 5.2 shows some examples of values of the speed of sound in common materials.

$$\longleftarrow \lambda \longrightarrow$$
Wavelength

Frequency (f) = number of waves passing a fixed point per second

Velocity = distance moved by a wave peak per second

$$= \lambda f$$

The relationship $v = \lambda f$ holds good for all wave motions

Fig. 5.1 The relationship between wavelength, frequency and velocity for a wave.

Material	Speed of sound (m/s)	
Aluminium	5100	
Copper	3700	
Glass	4500	(depending on constitution of glass)
Sea water	1510	
Air (20°C)	343	
Helium	965	
Hydrogen	1284	

Fig. 5.2 The velocity of sound in different materials.

The perception of sound by the ear is a much more complicated business. Objective measurements of sound waves can make use of the intensity, measured as the number of watts of sound energy per square metre of receiving surface, or of the wave quantities of pressure amplitude or displacement amplitude. The ear has a non-linear response, and a sensitivity that varies very markedly with the frequency of the sound. The response of the ear is at its maximum to sounds in the region of 2 KHz, as the threshold curve of Fig. 5.3 illustrates. This curve shows the amount of sound, measured in the three main ways, which is just discernible by the average human ear. The difference in the threshold intensity from the peak sensitivity frequency of about 2 KHz to a bass frequency of 100 Hz is very large, a ratio of about 10^5 in terms of watts per square metre.

There is less variation in the threshold of pain (Fig. 5.4), but this curve also shows the greater sensitivity to sound in the 2 KHz region. The pain threshold

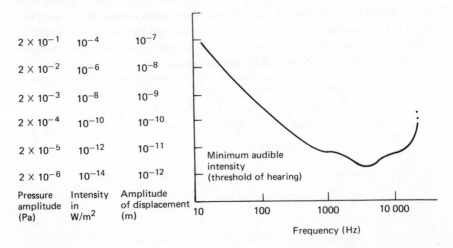

Fig. 5.3 The minimum detectable sound level for the average ear plotted for different frequencies.

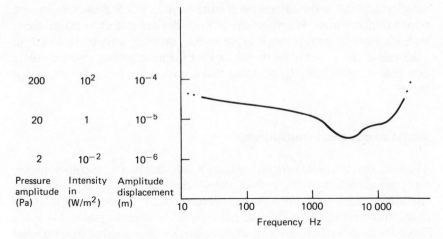

Pressure amplitude (Pa)	Intensity in (W/m²)	Amplitude displacement (m)
200	10^2	10^{-4}
20	1	10^{-5}
2	10^{-2}	10^{-6}

Fig. 5.4 The pain level for the ear, plotted for the frequency range. This shows much less variation with frequency.

is approached much more frequently in the latter part of the twentieth century than in earlier times, and frequent bombardment of the ears to this threshold results in deafness. Legislation for health and safety at work is now having some impact on deafness caused by noisy machinery, but nothing seems to put an end to the excessive amplification used in the entertainment industry which is responsible for a much greater incidence of hearing problems. This will presumably right itself when audiences learn that they can get better acoustic conditions by paying nothing and remaining outside the hall.

The frequency range over which sound can be detected by the human ear is limited to the range of about 20 Hz to 20 KHz, and in old age the upper limit is reduced on average by about 1 Hz per day. The lower limit is determined by the sound-filtering effect of tissues in the ear, and avoids the unpleasant effect of the many low-frequency vibrations that exist around us. The transducers that are the subject of this chapter, however, are not necessarily constrained to these frequency limits, and some can be used with infrasound (very low frequencies) or with ultrasound (very high frequencies). Acoustic waves, in fact, can make use of frequencies in the MHz range, well above the upper limit of audible sound, and in this region waves become much more directional and much more subject to filtering effects that can be achieved by shaping the wave path. The topic of surface acoustic filters is based on the effects of surface shape on the transmission of acoustic waves (SAW) along a (solid) surface.

The effect of a sound wave on a material is to vibrate the material, and in the course of this vibration every part of the material will be accelerated. The acceleration is in alternate directions, and there is no bodily displacement of the material, but an electrical output can be obtained from an accelerometer connected to the material. The sensors and transducers for sound to electricity are therefore of the same form as the transducers for acceleration and velocity, and the main differences are the ways in which these sensors and transducers are used. We make use of these devices mainly as transducers, since the aim of a microphone is to produce an electrical wave that is a faithful analogue of the sound wave that is striking the microphone, and the power conversion, rather than the detecting that a wave is present, is the important factor.

Audio to electrical transducers

The sound to electrical energy transducer is the microphone, and microphone types are classified by the type of transducer they use. In addition to the transducer, however, a microphone will use acoustic filters, passages whose shape and dimensions modify the response of the overall system. These are needed because each transducer will have its own response that is determined by resonances in the materials as well as by the transducing principle itself, and the correction to a more uniform response has to be made by means of the acoustic passages in the microphone housing. This type of compensation is preferable to the use of electrical methods, because acoustic filters can have much sharper effects with less impact on the rest of the frequency range.

The characteristics of a microphone are both acoustic and electrical. The overall sensitivity is expressed as millivolts or microvolts of electrical output per unit intensity of sound wave, or in terms of the acceleration produced by the sound wave. In addition, though, the impedance of the microphone is of considerable importance. A microphone with high impedance usually has a fairly high electrical output, but the high impedance makes it very susceptible to hum pickup, either magnetically or electrostatically coupled. A low impedance is usually associated with very low output, but hum pickup is almost negligible.

Another factor of importance is whether the microphone is directional or omni-directional. If the operating principle of the microphone is the sensing of the pressure of the sound wave, then the microphone will be omni-directional, picking up sound arriving from any direction. If the microphone responds to the velocity (speed *and* direction) of the sound wave, then it is a directional microphone, and the sensitivity has to be measured in terms of direction as well as amplitude of sound wave. The microphone types are known as pressure or velocity operated, omni-directional or in some form of directional response (such as cardioid).

The type of transducer does not necessarily determine the operating principle as velocity or pressure, because the acoustic construction of the microphone is usually a more important factor. If, for example, the microphone uses a sealed capsule construction, then the pressure of the sound wave will be the factor that determines the response. If the microphone uses a diaphragm or other moving element which is exposed to the sound wave on all sides, then the system will be velocity operated.

The carbon microphone

The carbon granule type of microphone was the first type to be developed for telephone use, and was still in use for telephones long after it had been abandoned for any other purposes, though it has now been replaced by the electret capacitor type (see later) even for telephone use. The principle is illustrated in Fig. 5.5, and uses loosely packed granules of carbon held between a diaphragm and a backplate. When the granules are compressed, the resistance between diaphragm and backplate drops considerably, and the vibration of the diaphragm can therefore be converted into variations of resistance of the granules. The microphone does not, therefore, generate a voltage and requires an external supply before it can be used.

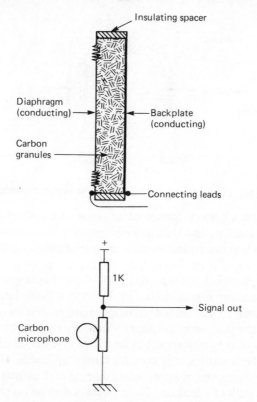

Fig. 5.5 The carbon granule microphone principle and the type of circuit into which the microphone is connected.

The sole advantage of the carbon granule microphone is that it provides an output which is colossal by microphone standards, with outputs of 1 V peak to peak possible. The linearity is very poor, the structure causes multiple resonances in the audio range, and the resistance of the granules alters in a random way even with no sound present, causing a high noise level. The

predominance of the carbon microphone in the early days of telephony was
due to its high output at a time when no amplification was possible, and the
introduction of valve and, later, transistor amplifiers caused the rapid demise
of the carbon microphone for serious audio use.

The moving iron (variable reluctance) microphone

A powerful magnet contains a soft-iron armature in its magnetic circuit, and
this armature is attached to a diaphragm. The principle is illustrated in Fig.
5.6. The magnetic reluctance of the circuit alters as the armature moves, and
this in turn alters the total magnetic flux in the magnetic circuit. A useful

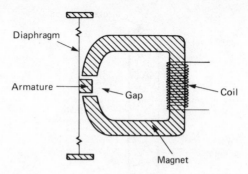

Fig. 5.6 The principle of the moving iron (variable reluctance) microphone.

comparison is of a battery connected in series with a fixed and a variable
resistor chain. A coil wound around the magnetic circuit at any point will give
an EMF which is proportional to each change of magnetic flux, so that the
electrical wave from the microphone is at 90 degrees phase to the sound wave
amplitude, proportional to the acceleration of the diaphragm. Most moving
iron microphones are manufactured in the form of sealed capsules or with
very limited access to one side of the diaphragm, so that the pressure of the
sound wave is the predominant quantity that determines the action.

The linearity of the conversion can be reasonable for small amplitudes of
movement of the armature, very poor for large amplitudes. The linearity can
be considerably improved by appropriate shaping of the armature and careful
attention to its path of vibration. These features depend on the maintenance
of close tolerances in the course of manufacturing the microphones, so that
there will inevitably be differences in linearity between samples of microphones of this type from the same production line.

The output level from a moving iron microphone can be high, of the order
of 50 mV, and the output impedance is also high, typically several hundred
ohms. Because the flux path in the transducer is almost closed, external
changes in magnetic field will be very efficiently picked up, and the result is
that the magnetic component of mains hum is superimposed on the output.

This can be reduced by shielding the magnetic circuit, using mu-metal or similar alloys. The magnetic circuit which is the predominant feature of this type of microphone also makes the instrument heavier than some other types.

Moving coil microphone

The moving coil microphone uses a constant-flux magnetic circuit in which the electrical output is generated by moving a small coil of wire in the magnetic circuit (Fig. 5.7). The coil is attached to a diaphragm, and the whole arrangement is usually in capsule form, making this pressure rather than velocity operated. As before, the maximum output occurs as the coil reaches maximum velocity between the peaks of the sound wave so that the electrical output is at 90 degrees phase angle to the sound wave.

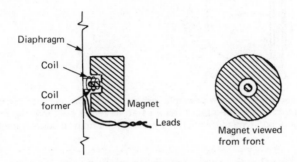

Fig. 5.7 The moving coil (dynamic) type of microphone has a coil wound on a former attached to the diaphragm and moving in an annular gap of a magnet.

The coil is usually small, and its range of movement very small, so that linearity is excellent for this type of microphone. The coil has a low impedance, and the output is correspondingly low, but not so low that it has to compete with the noise level of an amplifier. The low inductance of the coil makes it much less susceptible to hum pickup from the magnetic field of the mains wiring, and it is possible to use hum-compensating (non-moving) coils, known as humbuckers, in the structure of the microphone to reduce hum further by adding an antiphase hum signal to the output of the main coil.

Ribbon microphone

The ribbon microphone is the logical conclusion of the moving coil principle, in which the coil has been reduced to a strip of conducting ribbon (Fig. 5.8), with the signal being taken from the ends of the ribbon. An intense magnetic field is used, so that the movement of the ribbon cuts across the maximum

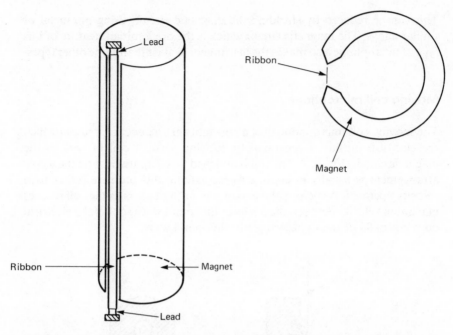

Fig. 5.8 The ribbon microphone principle, which is ideal for velocity operation. The sensitivity is very low, and the output requires an amplifier built into the microphone housing, but the ribbon type of microphone provides the highest sound quality.

possible magnetic flux to generate an electrical output whose peak value is as usual at 90 degrees phase to the sound wave.

One of several features that make the ribbon microphone unique is the fact that it is a velocity operated microphone, because the ribbon is affected by the velocity of the air in the sound wave rather than its pressure. This type of microphone is therefore used in situations where directional response is important, such as a voice commentary in noisy surroundings. The linearity is excellent, and the ribbon microphone is predominant where high quality of reproduction is of paramount importance.

The construction of the ribbon microphone inevitably makes the output extremely low, and the microphones are usually equipped with a transformer to raise both the signal voltage level and the impedance level. The hum pickup is also extremely low, and advantage can be taken of this to use a balanced output transformer to minimise hum pickup in the microphone leads (Fig. 5.9).

To be effective, a ribbon microphone has to be constructed to very precise limits, and good ribbon microphones are very expensive items, costing more than most domestic hi-fi users would contemplate spending on a complete sound system. The directional qualities are ideally suited to stereo broadcasting, though for some purposes the very directional response can be undesirable, and moving coil units have to be used.

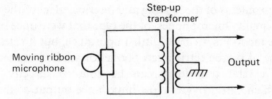

Fig. 5.9 The use of a step-up transformer which also provides a balanced output as a match between a ribbon microphone and a line.

Piezoelectric microphones

The piezoelectric transducer has the advantage over all the other types mentioned in this chapter of not being confined to use in air. A piezoelectric transducer can be bonded to a solid or immersed in a non-conducting liquid so as to pick up sound signals in any of these carriers. In addition, the piezoelectric transducer can be used easily at ultrasonic frequencies, with some types being usable into the high MHz region. All piezoelectric transducers require a crystalline material in which the ions of the crystal are displaced in an asymmetrical way when the crystal is strained. The linearity can vary considerably with the type of material that is used, and from sample to sample.

The original types of crystal microphones used Rochelle salt crystal coupled to a diaphragm. This ensured very high output levels (of the order of 100 mV), with very high output impedance and very poor linearity. Rochelle salt was not used for long because of its unfortunate habit of changing to an inactive form when kept at a moderately high temperature and humidity, and many pioneer users of tape recorders were puzzled to find that their microphones (and gramophone pickups) refused to work after the machine was brought down from the loft or up from the cellar after a hot summer.

The types of piezoelectric transducers that are used nowadays are mostly synthetic rather than natural crystals. One such material is barium titanate, which is used in piezoelectric transducers for frequencies up to several hundred kHz. The original type of piezoelectric microphone which used a diaphragm coupled to the crystal is seldom seen nowadays, because the sensitivity of modern piezoelectric materials to vibration is such that the impact of the sound wave on the crystal alone is enough to provide an adequate output. Most microphones of this type are made as pressure-operated types because one side of the crystal is normally used for securing the assembly to its casing.

The piezoelectric microphone has a very high impedance level and a much higher output than other types. The impedance level is of the order of several megohms, as distinct to a few ohms for a moving coil type. At this very high impedance level, electrostatic pickup of hum is almost impossible to avoid, along with the problems of the loading and filtering effect of the microphone cable. For low-quality microphones, of the type that were once supplied with tape or cassette recorders, this is of little importance, but it rules out the use of a simple type of piezoelectric microphone for studio purposes. For such purposes, the crystal transducer can be coupled directly to a MOS preamplifier which can provide a low-impedance output at the same high voltage level as is provided by the piezoelectric transducer. The preamplifier operating voltage can be supplied from a built-in battery to avoid the problems of running supply cables along with signal cables.

Capacitor microphones

The capacitor microphone is a remarkable example of a principle that was comparatively neglected until another equally old idea was harnessed along with it. The outline of a capacitor microphone is illustrated in Fig. 5.10. The amount of electric charge between two surfaces is fixed, and one of the surfaces is a diaphragm which can be vibrated by a sound wave. The vibration causes a variation of capacitance which, because of the fixed charge, causes in turn a voltage wave. The output impedance is very high, and the amount of

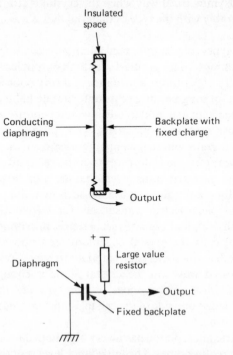

Fig. 5.10 The capacitor microphone principle. The conducting diaphragm is earthed and the backplate is fed through a high value resistor of several megohms so as to produce an approximation to constant charge conditions.

output depends on the normal spacing between the plates – the smaller this spacing, the greater the output for a given amplitude of sound wave. The construction of the microphone ensures that it is always pressure operated.

The two main objections to the capacitor microphone in the past were the need for a high-voltage supply and the hum pickup problems of the very high impedance. The high-voltage supply (called a polarising voltage) was needed to provide the fixed charge; this was done by connecting the supply voltage to one plate through a very large value resistor. The high impedance made it difficult to use the microphone with more than a short length of cable (which added to the 'dead' capacitance), though it was suited to the valve input stages of the time.

The capacitor microphone can be very linear in operation and can provide very good quality audio signals without the need for elaborate constructional techniques. This was realised by a few users, notably Grundig, who always provided capacitor microphones with their tape recorders in the early days of domestic tape recording. There were also a few manufacturers who specialised in high-quality capacitor microphones for studio use, but capacitor microphones were always a rarity in comparison with moving iron and moving coil types.

The revival of the capacitor microphone came about as a result of a revival of interest in an old idea, the electret. An electret is the electrostatic equivalent of a magnet, a piece of insulating material which is permanently charged. The principle has been known for a century, that if a hot plastic material (in the broadest sense of a material that can easily be softened by heating) is subject to a strong electric field as it hardens, it will retain a charge for as long as it remains solid. Materials such as acrylics (like Perspex) are electrets, and the idea was once considered for the manufacture of monoscope tubes, tubes which provide a TV test signal.

A slab of electret, however, is the perfect base for a capacitor microphone, providing the fixed charge that is required without the need for a polarising voltage supply. This allows very simple construction of a capacitor microphone, consisting only of a slab of electret metallised on the back, a metal (or metallised plastic) diaphragm, and a spacer ring (Fig. 5.11), with the

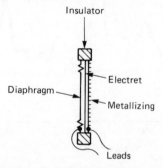

Fig. 5.11 The electret capacitor microphone, which can be manufactured in very small sizes and which needs no polarising supply.

connections taken to the conducting surface of the diaphragm and of the electret. This is now the type of microphone which is built into cassette recorders, and even in its simplest and cheapest versions is of considerably better audio quality than the piezoelectric types that it displaced.

The high impedance of the capacitor electret microphone is no handicap now that MOS preamplifiers can be used, and for studio quality capacitor microphones the preamplifier can be battery operated so as to avoid the need for supply lines. The linearity of the microphone is independent of the electret provided that the electret can supply a truly fixed charge. If the electret becomes leaky, as is possible through surface contamination, then the low-frequency response of the microphone will be degraded.

Microphone problems

Each specific type of microphone has its own problems and advantages, but there are problems which are common to all microphone types. The main problem of this kind is resonance, which will cause the output from the microphone to be distorted at some frequency, to form either a peak or a trough (Fig. 5.12). These resonances can be electrical, but are much more likely to be mechanical and, as such, more difficult to deal with.

The two main techniques for dealing with mechanical resonances are shifting and damping. A resonance can be shifted by altering the mass of the resonating part, so that the resonance occurs outside the audio region. Reducing the vibrating mass will have the effect of shifting the resonance to a higher frequency, and when this technique is used, the aim is usually to shift

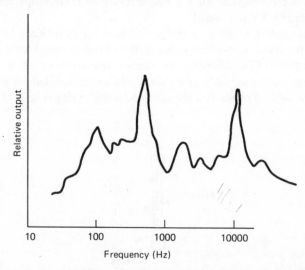

Fig. 5.12 The type of response that can be obtained from an uncorrected microphone element. The peaks and troughs are due to resonances, mechanical and electrical.

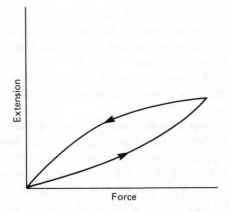

Fig. 5.13 The extension/force characteristics for a typical high-hysteresis rubber material. The area enclosed by the loop is a measure of the amount of damping that can be obtained.

the resonance to 30 KHz or higher. When the resonance is (unusually) at a low audio frequency, then adding mass can shift it to a lower, sub-audio frequency – this is more often a loudspeaker problem than a microphone problem.

Damping a resonance means that the energy of the resonating material must be dissipated, and this has to be done by using yielding materials rather then elastic materials like metals. Synthetic rubber materials can be made which have very high mechanical hysteresis (Fig. 5.13) and which make excellent damping materials to support diaphragms and other vibrating parts, and in general the use of damping calls for a considerable knowledge of materials of this type. Damping alone is seldom a cure for a bad resonance, and design effort has to be directed both to frequency-shifting the resonance and to damping it.

Electrical to audio transducers

The microphone types that we have dealt with would be of little use unless we also had transducers for the opposite direction, and such transducers have been used for even longer than microphones. Earphones were used for electric telegraphs in which the transmitter consisted of a Morse key, so that the earphone predated the microphone by a considerable number of years. The use of both earphone and microphone in a system is attributed to Bell, who in trying to develop a hearing aid for the deaf (a passionate interest that he shared with his father, and a cause that has benefited greatly from the Bell fortune) created the first working telephone system – and also laid the foundation for a telephone system in the USA which makes that of any other country seem pitifully inadequate.

Until the use of thermionic valves became common in radio receivers, loudspeakers were a comparatively rare sight, though the basic principles had existed for some considerable time. Without power amplification, however, the use of a loudspeaker was pointless, and this was the main reason magnetic recording lagged so much behind disc recording. The output from the mechanical disc gramophone was from a form of (non-electrical) loudspeaker and could be heard over a whole room, whereas the output of the early steel-wire recorders (Poulsen's Telegraphone) could be heard only on earphones.

For each type of microphone, there is a corresponding earphone and loudspeaker type, and in the following section we shall look at the various types in detail. Since the basic principles of most of them have been considered under the heading of microphones, these will not be repeated and we shall concentrate on the features that are unique to the purpose of transducing electrical signals into audio waves.

Of the two, the task of the earphone is very much simpler, and the construction of an earphone that can provide an acceptable quality of sound is very much simpler (and correspondingly cheaper) than that of a loudspeaker, as any Walkman addict will confirm. The earphone can use a small diaphragm, and ensure that the sound waves from this diaphragm are coupled directly to the ear cavity. The power that is required is in the low milliwatt level, and even a few milliwatts can produce considerable pressure amplitude at the eardrum – often more than is safe for the hearing. The conditions of use are, in other words, strictly defined, and the designer can concentrate on reducing resonances and increasing linearity in the certain knowledge that the effect will be noticeable.

The loudspeaker designer has a much more uphill task. A loudspeaker is not used pressed against the ear, so that the sound waves will be launched into a space whose properties are unknown. In addition, the loudspeaker cannot be used alone but has to be housed in a cabinet whose resonances, dimensions and shape will considerably modify the performance of the loudspeaker unit. The assembly of loudspeaker and cabinet will be placed in a room whose dimensions and furnishing are outside the control of the loudspeaker designer, so that a whole new set of resonances and the presence of damping material must be considered. One designer has said that if we had not known loudspeakers before now, it's unlikely that we would even consider trying to make them today, just as we would have banned the use of steam if environmentalists had been around in the nineteenth century. Fortunately, human beings show a marked reluctance to return to caves after tasting better things.

The transducer of a loudspeaker system is sometimes termed the 'pressure unit', and its task is to transform an electrical wave, which can be of a very complex shape, into an air-pressure wave of the same waveform. To do this, the unit requires a motor unit, transforming electrical waves into vibration, and a diaphragm which will move sufficient air to make the effect audible. The diaphragm is one of the main problems of loudspeaker design, because it must be very stiff, very light and free of resonances – an impossible combination of virtues. Practically every material known has been used for loudspeaker diaphragms at some time, from the classic varnished paper to titanium alloy and carbon fibre, and almost every shape variation on the traditional cone has been used.

The main cone problem is breakup. If the cone is to be able to handle low frequencies, it must have a large area. At high frequencies, however, there will be waves on the cone itself, so that different parts of the cone move in different directions (Fig. 5.14), causing the waves from the different parts of the cone to interfere with each other, and considerably modifying the response. The

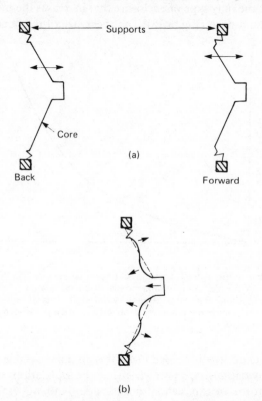

Fig. 5.14 The ideal cone movement (a) preserves the shape of the cone so that the whole cone moves in unison. Since a cone cannot be perfectly rigid, however, breakup occurs (b) in which different parts of the cone move independently, with the outer parts of the cone lagging behind the movement of the inner part because the driving force is located at the inner part.

usual solution to this problem is to use more than one driver unit and divide the electrical signal into low and high frequency components fed to the appropriate units. This is, after all, reasonably justified because few musical instruments produce a full range of sound frequencies and there is no reason why a single loudspeaker should. The low-frequency units are known as woofers and the upper frequency units as tweeters, but this attractive terminology is abandoned in the mid range where the units are simply termed mid-frequency units. The few loudspeaker designs that have achieved high-quality results with a single diaphragm, however, are among the best known.

The efficiency of loudspeakers is notoriously low, around 1%, mainly because of the acoustic impedance matching problem. In simple terms, most loudspeakers move a small amount of air with a comparatively large amplitude, whereas to produce a sound wave effectively they ought to move a very large amount of air at a comparatively low amplitude. This mismatch can be remedied to some extent by housing the loudspeaker in a suitable enclosure, but the only type of enclosure that increases the overall efficiency is the exponential horn (Fig. 5.15). The sheer size of the horn and the rigid

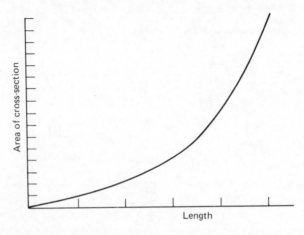

Fig. 5.15 The ideal graph of area of cross section plotted against distance from the driver unit for a horn. A horn of ideal shape needs to be very large in order to make its dimensions comparable with the largest wavelength that it will work with, and compromises such as folding the horn back on itself or driving the horn partway along its length have been used.

dense construction that is needed make this an unacceptable solution for all but the most avid listeners, one of whom had the foundations of his house cast so that they form an exponential horn shape. Nothing less really suffices, because it is only when the listening end of a horn is about 16 feet square that the benefits of horn-loading become really noticeable. The effect is still noticeable even for small linear horns, however, and the astonishing effect of bringing up the small end of a horn to a Walkman earphone is something that has to be heard to be believed.

The moving-iron transducer

The first type of earphone, as applied to the early telephones, was a moving-iron type, and this principle has been extensively used ever since. As applied

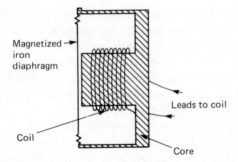

Fig. 5.16 The telephone earpiece principle. The magnetised diaphragm is made from iron or a magnetic alloy and is moved by the attraction or repulsion of the core as the signal current flows in the coil. The earpiece is very sensitive, though the sound quality is low.

to the telephone, the earphone uses a magnetised metal diaphragm (Fig. 5.16) so that the variation of magnetisation of the fixed coil will ensure the correct movement of the diaphragm. Without this fixed magnetic polarisation, the diaphragm frequency would be twice the frequency of the electrical signal. Earphones of this type are very sensitive, but the sound quality is very poor because of the comparatively stiff diaphragm which causes unavoidable resonances.

Miniature moving-iron earphones are still in use, particularly where sound quality is not of the highest importance, but the principle has died out as far as loudspeakers are concerned. At one time, moving-iron loudspeakers, which were virtually earphone units with a cone attached, were used, but these were soon replaced by the moving-coil type which provided noticeably better sound quality. The sound of the early loudspeakers has been aptly described as a 'mellow bellow', and this is no longer acceptable except inside cars.

The moving-coil transducer

The moving-coil principle as applied to loudspeakers and earphones has been extensively used, and the vast majority of loudspeakers use this principle. The use of moving-coil earphones has been less common in the past, but these are now in widespread use thanks to the miniature cassette player vogue. As applied to earphones, moving-coil construction permits good linearity and controllable resonances, since the amount of vibration is very small and the moving-coil unit is light and can use a diaphragm of almost any suitable material.

The use of moving-coil drivers with cones to form loudspeakers is by no means so simple. One problem of all loudspeakers is that the unit which reproduces the low frequencies needs a very freely suspended cone and must be able to reproduce large amplitudes of movement. This amplitude of movement can be a centimetre or more, and it is very difficult to ensure that the magnetic flux density around the moving coil is uniform over this distance. The design of the magnetic circuit calls for a large magnet and a large core, and a shape which is computed to provide the best available linearity of flux density. In addition, there is a conflict between making the coil and cone very freely suspended and yet maintaining the position of the coil concentric in the core of the magnet.

The magnets of modern moving-coil units are invariably permanent magnets, often using quite exotic alloys. At one time, moving-coil loudspeakers, known as 'dynamic' loudspeakers, use an electromagnet to provide the magnetic field, but this vogue was short-lived because of the demands it made on the power supply of the radio receiver that used it.

At the other end of the frequency scale, there is no requirement for large amplitudes of movement, and the main problems are of resonance and cone breakup. Metal cones are often used with success, but a large number of successful designs of loudspeaker systems use moving coil units only for the lower and mid frequencies, with other types used for the highest frequencies.

A variation of the moving-coil principle that has been successfully used for earphones is the electrodynamic (or orthodynamic) principle. This uses a diaphragm which has a coil built in, using printed circuit board techniques. The coil can be a simple spiral design, or a more complicated shape (for better linearity), and the advantage of the method is that the driving force is more evenly distributed over the surface of the diaphragm. This avoids breakup, and allows the use of a much more flexible diaphragm than would be possible otherwise. Headphones based on this principle have been very successful and of excellent quality.

Ribbon loudspeakers

The ribbon principle, already dealt with under the heading of microphones, is also used to provide loudspeaker action. The moving element of a ribbon loudspeaker is necessarily small, and for that reason, the unit is a tweeter rather then a full-range type. The ribbon construction, however, offers a very directional response and can be built into a small horn type of enclosure (Fig. 5.17), which makes it a very efficient transducer compared to others. Ribbons of about 5 cm long, corrugated along the length for stiffness, are used, and because the impedance is very low – a fraction of an ohm – matching transformers are usually built into the loudspeaker. With conventional construction, the ribbon tweeter can handle frequencies from 5 KHz upwards, but more elaborate design can allow this range to be extended down to about 1 KHz.

Wide-range multi-ribbon units are also feasible, but in a very different size (and price) category. The commercially available types use three units, of

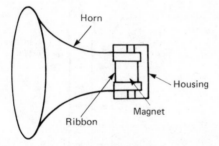

Fig. 5.17 The ribbon type of loudspeaker is usually combined with a small horn enclosure because the small size of the ribbon alone is insufficient to produce a large output.

which the bass unit is very large, and requires its own amplifier to supply about 100 W driving power.

Piezoelectric loudspeakers

The piezoelectric principle has been used in the manufacture of tweeters, for which it is reasonably suited if the problems of resonance can be dealt with.

The piezoelectric principle has also been used for earphones in the form of piezoelectric (more correctly pyroelectric, since the electrical parameters are temperature sensitive) plastics sheets which can be formed into very flexible diaphragms. The effect of applying a voltage between the faces of such a diaphragm is to make the dimensions shrink and expand as a waveform is applied, and this can be converted into a movement that will move air by shaping the diaphragm as part of the surface of a sphere. This can be done, for example, simply by stretching the material over a piece of spherical plastic foam. The moving mass is very small and sensitivity is high, with no need for a power supply. The linearity that can be obtained depends on the surface shape as well as on the piezoelectric characteristics.

Capacitor transducers

The possibility of constructing earphones or loudspeakers along the lines of a capacitor microphone has existed for a long time, but the practical difficulties have been satisfactorily resolved by only two designs, the Quad and the Magna-Planar wide-range electrostatic loudspeakers (Fig. 5.18). The main problem initially was that using a single-ended design analogous to a capacitor microphone provides very poor linearity, and the breakthrough was the principle, due to Peter Walker, of a diaphragm which had signal voltage applied to it and which faced a sheet of conducting material at a high voltage on each side. The system has also been used in earphones which, despite the need to provide a high polarising voltage for the plates, have been

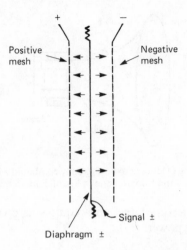

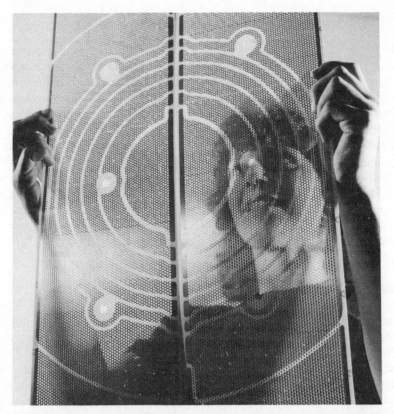

Fig. 5.18 The electrostatic loudspeaker principle (a). The signal voltage applied to the conducting diaphragm will provide charge, and the action of the field provided by the static metal mesh on each side will move the diaphragm. The important feature is that the whole diaphragm is moved, not simply a small portion. The world famous Quad loudspeakers (b) make use of this type of action.

very popular and of outstanding audio quality.

The advantage that makes the electrostatic loudspeaker principle so attractive is that the driving effort is not applied at a point in the centre of a cone or diaphragm, but to the whole of a surface that can be large in area. There is therefore no breakup problem, since all parts of the diaphragm are driven, and so a single unit can handle the whole of the audio range. Resonances can present a negligible problem, due to the overall drive, and no special enclosure other than is required for electrical safety and mechanical rigidity will be necessary. The original Quad design has now been in production for over 21 years and, though recently redesigned, still follows the original principles and provides outstanding quality, particularly for concert-hall performances.

The most recent design uses the 'point-source' principle, achieved by driving the diaphragm as a set of concentric circles which are not in phase. In this way (Fig. 5.19), the sound wave that is created appears to come from a point behind the diaphragm, and the practical effect of this is to make the sound appear to be independent of the loudspeaker in a way that is quite remarkable as compared to moving-coil units.

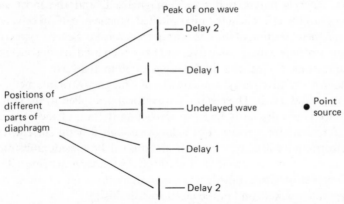

Fig. 5.19 The most recent Quad design divides the diaphragm into concentric circles, each driven separately with a small time delay. The effect of this is to make the diaphragm move as if it had been driven from a point source of sound.

The electret principle has not been used to date for loudspeakers, but earphones have been constructed which follow the basic capacitor type of design but using an electret to provide the fixed charge. These are claimed to give a performance that approaches that of the best electrostatic types, but at a much lower cost and free of the need to supply a polarising voltage.

Ultrasonic transducers

Though loudspeakers and microphones may use similar operating principles, the differences between receiving a sound wave and generating it are sufficient to make the actions interchange only very poorly. A loudspeaker can generally be used as a microphone, but as an insensitive and low-quality microphone. A microphone can be used as an earphone with some success, but is not suitable as a loudspeaker because it is not designed to handle the amount of power that is needed.

By contrast, the transducers that we use for ultrasonic waves are almost totally reversible. The transducers that are used for sending or receiving ultrasonic signals through solids or liquids can operate in either direction if required, but for ultrasonic signals sent through the air (or other gases), the transducers are used with diaphragms and in enclosures that can make the application more specialised, so that a transmitter or a receiver unit has to be used for its specific purpose.

The important ultrasonic transducers are all piezoelectric or magnetostrictive, because these types of transducers make use of vibration in the bulk of the material, as distinct from vibrating a motor unit which then has to be coupled to another material. The magnetostrictive principle has not been mentioned before, because it is not generally applied in the audio frequency range. Magnetostriction is the change of dimensions of a magnetic material as it is magnetised and demagnetised, and the most common experience of the effect is in the high-pitched whistle of a TV receiver, caused by the magnetostriction of the line output transformer. Several types of nickel alloys are strongly magnetostrictive, and have been used in transducers for the lower ultrasonic frequencies, in the range of 30 to 100 KHz.

A magnetostrictive transducer consists of a magnetostrictive metal core on which is wound a coil. The electrical waveform is applied to the coil, whose inductance is usually fairly high, so restricting the use of the system to the lower ultrasonic frequencies. For a large enough driving current, the core magnetostriction will cause vibration, and this will be considerably intensified if the size of the core is such that mechanical resonance is achieved. The main use of magnetostrictive transducers has been in ultrasonic cleaning baths, as used by watchmakers and in the electronics industry.

The piezoelectric transducers have a much larger range of application, though the power output cannot approach that of a magnetostrictive unit. The transducer crystals are barium titanate or quartz, and these are cut so as to produce the maximum vibration output or sensitivity in a given direction. The crystals are metallised on opposite faces to provide the electrical contacts, and can then be used either as transmitters or as receivers of ultrasonic waves. The impedance levels are high, and the signal levels will be millivolts when used as a receiver, a few volts when used as a transmitter.

Though piezoelectric transducers are used in ultrasonic cleaners, their main applications are in security devices and signal processing. An ultrasonic transmitter in air can 'fill' a space (such as a room or a yard) with a standing-wave pattern, and a receiver can detect any change in this pattern that will be caused by any new object. Ultrasonic units cannot necessarily distinguish between a cat and a cat-burglar, however. For signal processing, a radio-frequency signal can be converted to an ultrasonic signal by one transducer and converted back by another, using the ultrasonic wave path, usually in a solid such as glass, to create a delay. This use of an ultrasonic delay is essential to the PAL colour TV system that is used in most of Western Europe. Ultrasound waves can also be filtered by mechanically shaping the solids along which they travel, and this is the basis of surface wave filters.

Infrasound

The wave frequencies below 20 Hz are not so extensively used as the ultrasonic frequencies, but the sensing of these frequencies is a matter of some importance. The vibrations of the Earth that are accompanied by earthquakes are in the very low frequency region, and are termed seismic waves. Transducers for seismic waves must be capable of very low frequency response, which rules out the use of piezoelectric transducers, and most seismic transducers work on the principle of using a suspended mass to operate a transducer of the capacitive or inductive type, such as the example in Fig. 5.20. The principle here is that the vibrations of the Earth will move the

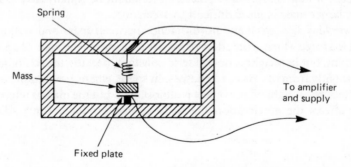

Fig. 5.20 Seismic detector principle. The mass remains still while the casing vibrates with the Earth. The relative movement is detected, in this case, by a capacitive transducer. The casing is evacuated to avoid the damping effect of the air.

casing, leaving the suspended mass at rest, and that the relative motion between the supports and the mass will produce the output. The other natural source of very low frequency vibrations is the communication of whales, and the signal strength is often enough to permit a piezoelectric type of transducer to be used, with a large diaphragm mechanically coupled to the piezoelectric crystal and the output from the crystal connected to a MOSFET amplifier, with a chopper stage used for the main amplification. The usual technique with these signals is to tape-record them with a slow-running tape, and replay at a higher speed to make it easier to display (and hear) the waveforms.

Chapter Six
Solids, Liquids and Gases

Mass and volume

Of all the primary physical quantities, mass is the most difficult to define in terms that mean anything to a non-physicist. The official definition is that mass is the quantity of matter in an object, but this is not exactly a step forward in understanding. A more useful, though less fundamental, idea of mass is as the ratio of force to acceleration when a force is applied to a mass. The greater the mass of an object, the less acceleration a given force can produce. Two different masses placed on ice might be equally easy to move, but the larger mass is more difficult to accelerate.

In everyday life, we make no difference between mass and weight, but weight is a force which is the effect of gravity on a mass. The mass of an object is constant, but its weight is not quite so constant, because the size of gravity varies slightly from one place to another. In space, where gravitation has very little effect, the weight of any object is almost zero, but the mass is unchanged and it affects the acceleration that a force can produce. The traditional

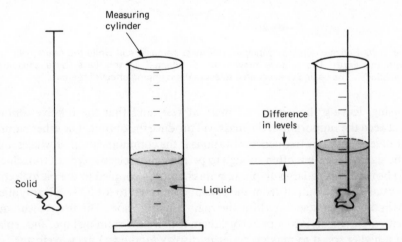

Fig. 6.1 Finding the volume of a porous or irregular solid by displacement. The volume of liquid displaced by the solid is equal to the volume of the solid unless the solid dissolves in the liquid.

method of measuring mass, as distinct from weight, is the mass balance, in which an unknown mass is balanced against a set of masses of known values. Because gravity affects both sides of the scale, this is a comparison of masses rather then weights, though it is not effective if the gravity is zero.

Volume is much easier to define, as the amount of space that an object takes up. Even this, however, can be deceptive. The volume of a slab of glass is easy to appreciate, but what is the volume of a porous material like a sponge? The volume of a solid non-porous material can be calculated from its dimensions, but for a porous material we have to measure volume by finding what volume of liquid the object will displace (Fig. 6.1). For a gas, the volume is the volume of the container because a gas has no fixed volume or shape.

Electronic sensors

The mass and volume of solids are quantities that are not easy to sense or measure by electronic methods. The main methods for mass really depend on a weight measurement, but for most purposes this is what is required in any case, since the distinction is academic when you are working on Earth. The available methods make use of strain gauges to measure weight, or position gauges to measure the extension of a spring by the weight of a mass.

Figure 6.2 shows the principle of a simple lever-arm weight sensor. The arm is made of an elastic material, in the sense of a material whose change of dimensions will be proportional to the force applied to it. The arm material will usually be metal whose cross-section will be chosen to suit the weights to be measured, which could be in the gram range or in the tonne range, depending on what is to be weighed. A strain gauge, or more likely a pair of strain gauges, is used to obtain an electrical output which is proportional to the amount by which the lever arm is bent. This in turn will be proportional to the weight applied at the end.

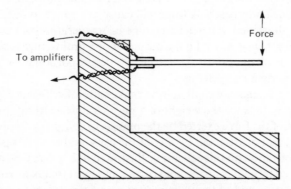

Fig. 6.2 The lever-arm weight sensor. The application of a force to the end of the arm causes the arm to bend, and the amount of bending is sensed by the strain gauges. The device requires calibration with known weights.

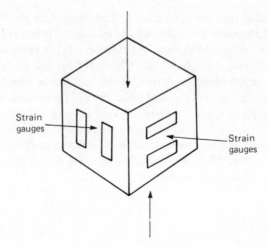

Strain gauges

Strain gauges

Fig. 6.3 A 'force cube' sensor, which makes use of an elastic material in cube form with strain gauges to detect changes of dimension caused by loading. The method is particularly suited to large forces, using a cube of metal rather than of more easily compressed materials.

Another form of strain-gauge weight sensor is illustrated in Fig. 6.3. This consists of a cube of elastic material on which strain gauges have been fastened so as to detect changes in the dimensions. Placing a weight on top of the cube, with the cube resting on a support, will distort the cube, giving a reading from the strain gauges that will once again be proportional to the amount of weight or force.

Both types of strain-gauge measurement need to use more than one strain gauge because of the effects of temperature. The effect of a small change of temperature on an elastic material is likely to cause more change of dimension than the effect of a modest force, so that some form of temperature compensation is needed. This usually takes the form of multiple strain gauges, and in the lever type of sensor is best achieved by using one gauge on top of the lever and one underneath. A force will cause one gauge to be lengthened and the other shortened, whereas a change of temperature will cause both gauges to be lengthened, so that if the output of one strain gauge is subtracted from the output of the other, the result will be proportional to force and unaffected by temperature changes.

A third form of mass/weight sensor makes use of a balance principle, though usually to balance the weight of a mass against a strong spring (Fig. 6.4). The effect of the force exerted by the mass is to compress the spring, and this movement can be detected by any of the sensors that were discussed in Chapter 2. A variation on this scheme is illustrated in Fig. 6.5. One end of the arm is supported by a light spring, and carries a magnet that is surrounded by a coil. The opposite end of the balance arm forms a capacitive detector, the output from which is amplified and used to provide a current that is circulated through the coil. When the application of a force tries to deflect the arm, the

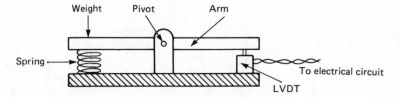

Fig. 6.4 A form of lever-arm spring balance in which the force applied to the arm causes compression of the spring and the amount of movement (proportional to force) is sensed by any suitable transducer, such as an LVDT. One advantage of this method is that by changing the position of the spring several ranges of force values can be measured.

deflection produces a signal at the capacitive sensor, and the output current acts on the magnet so as to restore the original position of the arm. Because of the very large negative feedback, the movement of the balance arm is always very small, so that the linearity of measurement can be good. The force need not be applied directly to the end of the arm, so that a large range of forces can be catered for by making the point of application of the force variable.

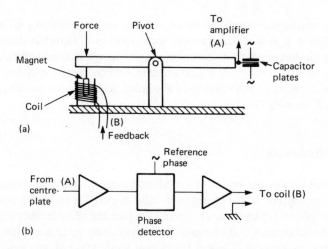

Fig. 6.5 The very sensitive feedback balance system (a) in which no springs are used. The electrical circuit (b) shows that the signal from the sensing capacitor is used to restore the balance arm position by applying current to the coil, and the amount of current that needs to be applied in this way is a measure of the force. Calibration is necessary, because the relationship between force and current is not easily calculated.

For tensile forces which are pulling forces, an adaptation of the basic strain-gauge design is illustrated in Fig. 6.6. This uses an elastic rod or tube which will be distorted when a tensile force is applied, causing an output from the strain gauges. As before, multiple strain gauges are used in order to reduce the effects of temperature changes.

The volume of a solid is not amenable to electronic measurement. The

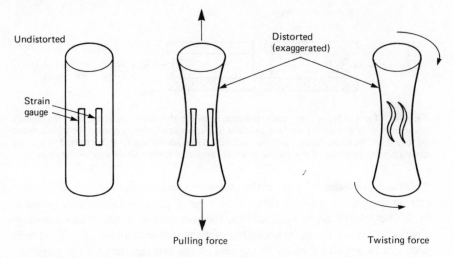

Undistorted

Distorted
(exaggerated)

Strain
gauge

Pulling force

Twisting force

Fig. 6.6 The 'elastic cylinder' method of sensing tensile or torsional forces, using a cylinder with strain gauges attached.

principle of displacement, however, allows indirect sensing of volume, because when a solid is completely immersed in a liquid it displaces an amount of liquid equal to its own volume. A change of liquid level, which is proportional to a change of liquid volume, can be sensed and this reading used as a measure of the volume of the solid. See later for a description of liquid level sensors.

Counting objects

The sensing of mass or volume of a solid is often much less important than the counting of objects. If, for example, a solid material is formed into units of known mass and volume, then counting the number of such units is a way of finding the total mass and volume, and is also likely to be a quantity that is needed for its own sake. The most common application of counting is to the number of objects passing a fixed point along a moving belt.

The simplest counting system consists of a light beam and sensor (see Chapter 3) coupled to an amplifier and a digital counter. The snag here is that if the system uses visible light, then the count can be upset by sudden changes in the ambient light level. This can sometimes be overcome by using a simple hood over the counting bay, and another method is to use infrared rather than visible light for the beam.

Ultrasonic beams can also be useful for counting purposes, but some care has to be taken over the wavelength of the sound wave, which must be much smaller than any dimension of the object or the distance between the objects. The danger here is that an interference pattern can be set up which will result in a false count. This is less likely if the ultrasonic transmitter produces a narrow-angle beam.

If the objects to be counted are made from magnetic material, the use of magnetic proximity methods, such as Hall-effect detectors, can often be advantageous. The counter can then operate in normal lighting, and the presence or absence of other objects, provided they are non-magnetic, will have no effect. If magnetic proximity methods are used, however, the distance between the objects and the detector has to be maintained more closely than would be the case for beam-interruption methods.

Liquid levels

A sample of liquid has a mass that is fixed, and a volume that changes slightly with changes of temperature. The shape of a liquid is, in general, the shape of its container, though in zero gravity most liquids take the shape of least surface area, a sphere. The measurement or sensing of the level of a liquid in a container is therefore a useful quantity which is proportional to the volume of the liquid.

The simplest form of liquid level gauge is based on the ancient float-and-arm system, as illustrated in Fig. 6.7. Any suitable transducer can be used to detect the angular movement of the arm, and a potentiometer is particularly useful. With a fixed voltage, AC or DC, applied to the fixed ends of the potentiometer, the voltage at the variable tap will be proportional to the position of the float between the low and high extremes.

The float system is less attractive if the liquid is corrosive. Though the float itself, and the arm, can be made from resistant material such as polypropylene, the transducer is still liable to damage from corrosive vapours, and a potentiometer is unlikely to be useful. The LVDT principle can be used, with the metal plunger encased in plastic and the wires insulated also with plastic, but other methods are usually preferable in such circumstances.

One other method is the air-compression sensor (Fig. 6.8). A tube is immersed in the liquid so that air (or possibly another liquid) is trapped inside the tube. A change in the liquid level outside the tube will cause the air to be compressed or expanded, and this change of pressure can be sensed. A scheme

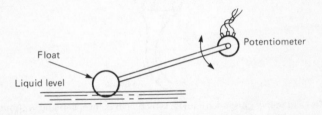

Fig. 6.7 The simplest method of liquid level sensing, using a float arm and a potentiometer.

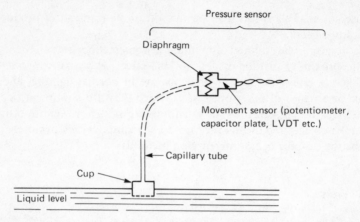

Fig. 6.8 A very common form of liquid level sensor which makes use of the liquid level to compress air. The air pressure is then sensed remotely, with a connection by way of a capillary tube.

like this is often used to sense water level in clotheswashers and dishwashers. For switching purposes, a diaphragm and microswitch can be used, but in order to sense the level the transducer should be a manometer or similar pressure detector (see Chapter 1).

A very neat method that can be applied to any liquid makes use of the permittivity of the liquid which will cause a change of capacitance if the liquid is placed between capacitor plates. The scheme is illustrated in Fig. 6.9,

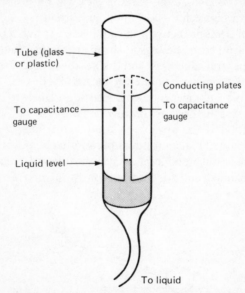

Fig. 6.9 The capacitive gauge for liquids, which allows level sensing without any direct contact with the liquid. The permittivity of the liquid is invariably greater than that of air, so if the liquid is a non-conductor, a change in level will cause a change in the capacitance between the plates.

showing the transducer as a pair of rectangular foil plates wrapped around a tube so that the edges do not quite touch. These plates now form a capacitor for which the content of the tube acts as a dielectric. Replacing the air in the tube by a liquid will cause a change of capacitance due to the different permittivity of the liquid, and the amount of the change will be proportional to the level of liquid in the tube between the limits of the length of the capacitor plates.

The change of capacitance is most easily sensed by making the capacitor part of an oscillating circuit, though some care is needed if the liquid is a partial conductor, as this can damp the oscillation and cause the oscillator to stop. The change of capacitance causes a change of oscillating frequency, with the frequency falling as the liquid rises, and this change of frequency can be converted into a voltage by a frequency discriminator or by a digital-analogue converter.

Liquid flow sensors

Liquid flow is a very important part of many manufacturing processes, and where the manufacturing process is continuous rather than a batch process the sensing and measurement of liquid flow is a vital part of plant control. The sensing of liquid flow can take three different forms. One is vector flow, in which the speed and direction of the liquid need to be sensed. Another possible requirement is volume flow, measuring the volume per second of liquid passing a point in a pipe. The third possible requirement is mass flow, which is usually calculated from volume flow, using the relationship that mass = volume × density. For most automation requirements, with liquids confined in tubes, the volume flow is the important one.

There is no simple and straightforward method of measuring liquid volume flow, though sensing flow is comparatively simple. One of the simplest methods is based on the pressure difference that occurs when the liquid flow is through an orifice, which can be an orifice plate or a nozzle (Fig. 6.10). For a liquid whose flow rate is slow enough to be laminar (streamline), as defined in Fig. 6.11, the pressure difference across the orifice is linearly proportional to the rate of flow, and this pressure difference can be sensed or measured by a manometer or by the difference in level between two tubes. If the flow is not laminar but is turbulent, the pressure is more likely to be proportional to the square of liquid speed.

If the presence of mechanically moving parts in the liquid stream is permissible, the 'mill-wheel' method can be used. In this method (Fig. 6.12), the movement of the liquid turns a turbine wheel which in turn is coupled to a tachometer. This method is quite acceptable for liquids whose flow rate is comparatively slow, and over a fair range of liquid velocity the output is reasonably proportional to the flow rate. The device needs calibration, however, either from another form of gauge or by the ancient method of

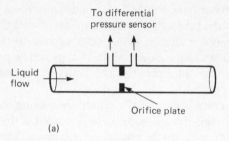

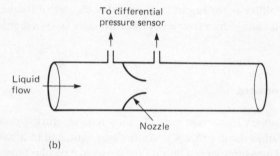

Fig. 6.10 Detecting liquid flow rate by means of the pressure difference across an aperture (a) or a nozzle (b). The pressure difference is proportional to rate of flow only while the rate is low (laminar flow), and the nozzle type usually permits a greater range of flow rates to be measured.

Laminar flow means that each layer of a liquid flows in a line which does not intersect with any other layer of a liquid. These lines are called stream lines, and laminar flow is sometimes known as stream line flow. When the rate of flow increases above a critical level, the flow becomes turbulent, with the stream lines breaking up into vortex paths. The resistance to flow then increases greatly.

The critical flow velocity is given by:

$$V_c = (RN)\frac{\eta}{\rho a}$$

RN = Reynolds number
η = viscosity of liquid
ρ = density of liquid
a = cross-sectional area of tube

Reynolds number is about 1000 for tubes of circular cross-section.

Fig. 6.11 Laminar and turbulent flow, and how critical velocity of flow is related to liquid viscosity, density and Reynolds number.

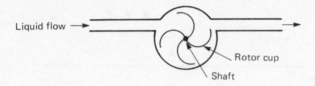

Fig. 6.12 The 'mill-wheel' method for liquid velocity sensing. The shaft can be coupled to any angular velocity sensor, such as a tachogenerator or a synchro.

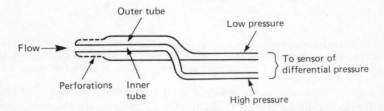

Fig. 6.13 The pitot-tube method of measuring gas flow, making use of the pressure difference between gas striking an orifice and gas passing over an orifice.

measuring how much liquid (volume or mass) is delivered in a given time. Another, more crude, method uses a vane mounted on a spring-loaded shaft and placed in the liquid flow, and senses the deflection of the vane by the angle through which the vane is turned. This angle is most simply sensed by a potentiometer connected to the shaft.

For sensing and measuring vector flow, the pitot tube is a method that has been used for some considerable time, particularly for the indication of aircraft air speed. The principle is illustrated in Fig. 6.13 in its most elementary form, consisting of a liquid U-tube manometer whose two ends are both fed from nearby points. One of these points is directed at the oncoming airstream, and the other is a perforated tube over which the airstream passes. The pressure difference between these two will depend on the speed and direction of the airstream. The pitot-tube principle can be used with gases and with liquids of low viscosity, and can be adopted for electronic sensing methods by using an electronic manometer.

A method that can be used only for conducting liquids is the magneto-hydrodynamic sensor, whose principle is illustrated in Fig. 6.14. Two electrodes are in contact with the moving liquid in a tube, and a strong magnetic field is applied at right angles both to the axis of the tube and to the line joining the electrodes. Motion of the liquid will cause an EMF to be generated between the electrodes, in accordance with the Faraday principles, with the amplitude of the EMF proportional to the rate at which the flux of the magnet is cut, so proportional to the velocity of the liquid.

The requirement for the liquid to be conducting makes this method of little

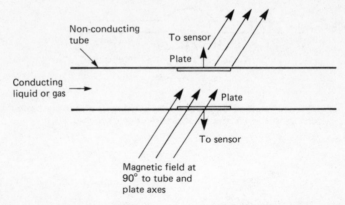

Fig. 6.14 The magnetohydrodynamic principle. The flow of the conducting fluid so as to cut the magnetic field causes an EMF to be generated between the plates, and the size of the EMF is proportional to the fluid velocity if the conductivity of the fluid is constant.

use for many applications, but it can be used with ionised gases, and has had some limited application as a transducer used to generate electricity from exhaust gases of both gas turbines and coal-fired generating systems.

Timing methods

Many sensors of liquid flow are based on the principle that affects waves moving through liquids or gases that may themselves be moving. The equations show how the velocity of a received wave from a stationary source can be affected by movement of the medium that carries the waves, and the relevant equation is illustrated in Fig. 6.15. The effect shows that if a wave is

Fluid velocity v →   Waves

$((((((\circ)))) \,) \,)^{\,)} \,) \,) \quad)$ • Receiver

Source →

← d →

If the normal velocity of sound is V_0 then the velocity downstream is $V_0 + V$

The time needed in still liquid is $t = \dfrac{d}{V_0}$

The time needed in moving liquid is $t = \dfrac{d}{V_0 + V}$ if a receiver is downstream,

or $t = \dfrac{d}{V_0 - V}$ if no receiver is upstream

Fig. 6.15 Detecting fluid movement by the time needed for a sound wave to travel between two points. This is a form of Doppler effect, but it does not cause any change of received frequency as happens when either source or receiver moves.

detected downstream of the moving liquid that carries the wave, then the time taken to reach the receiver is less than it would be in a liquid at rest, and if the wave is moving upstream then the time needed to reach the receiver is greater than it would be in a liquid at rest.

The waves that can be used for measurement can be ultrasonic, and ultrasonic flow gauges have been commercially available for a considerable time, ever since small ultrasonic transducers became available. Figure 6.16 shows the principle of the method, using in this example the simpler system of a receiver transducer downstream of the transmitter. This makes the velocity of the received wave equal to the natural velocity (in a still liquid) plus the liquid velocity, so that the change in wavelength is proportional to liquid velocity, and the absolute value of liquid velocity can be calculated with relative ease.

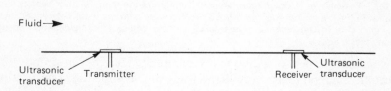

Fig. 6.16 Using ultrasonic transducers to measure fluid flow. The method is complicated by the effect of reflections from the pipe, and careful calibration is needed.

Ultrasonic methods are most useful when they can be applied in large-diameter tubes, using transducers beaming in the line of the liquid flow. Problems arise when the ultrasonic waves are reflected from the sides of the tube or from bubbles or cavities in the liquid, or when the ultrasonic transducers have to be used beaming across the direction of liquid flow. This latter method is, unfortunately, the most practical from the point of view of measuring the liquid flow with nothing immersed in the liquid. For many types of measurement, however, the cross-beam method is completely satisfactory, depending on the liquid and the shape of the cross-section of the tube.

Laser Doppler methods are also in use now, though they depend to some extent on the use of transparent tubes, or tubes with transparent insets, and on the liquid itself being transparent to the laser frequency. One method is illustrated in Fig. 6.17, in which the scattering of the light by tiny discontinuities in the liquid (bubbles, for example) is used to cause interference between two parts of a laser beam, causing interference and a frequency shift of the scattered beam. The Doppler laws as applied to light obey the principles of relativity inasmuch as the velocity of the light is constant throughout a medium, irrespective of the movement of the medium. This means that the frequency shifts that are observed are caused by changes of the light wavelength, never by a change of light velocity. A laser fluid velocity meter based on these principles can operate with liquid velocities of

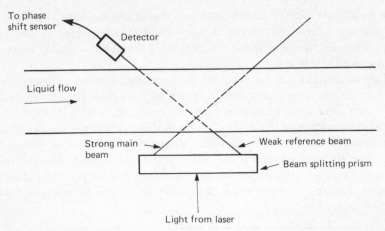

Fig. 6.17 Principle of Doppler laser sensor. The light of the main beam is scattered by particles in the liquid, and this scattered light interferes with the weak reference beam, causing a shift in the received interference pattern whose amount depends on the velocity of the liquid.

a few millimetres per second up to several hundred metres per second.

Another method that is considerably less simple is the use of correlation techniques. The sensor system consists of a laser beam which is split so that it can be reflected from two different points in the liquid, one downstream of the other. The received signals that are obtained should be identical if one is delayed relative to the other with a delay equal to the time taken for the liquid to traverse the distance between the beams – Fig. 6.18 shows the principle as

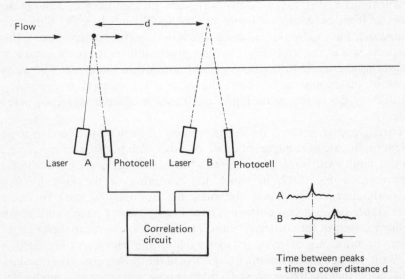

Fig. 6.18 The correlation method, which depends on sensing the time difference between identical changes in the reflected beam. Two lasers have been shown for simplicity, but in practice a single laser would be used along with prisms or optical fibres.

applied to one reflecting particle in the liquid. The use of the technique depends on electronic methods of correlating the readings, and digital circuits consisting of exclusive-OR gates are used. For calibration, bubbles may have to be injected into the liquid, but most liquids will provide sufficient natural discontinuities to make the system workable.

Gases

The velocity of gases can be measured in the same ways as are used for liquids, but a few specialised methods are added. The anemometer is used for wind-speed measurements in meteorology, and consists of a set of cup-shaped vanes which rotate in a horizontal plane (Fig. 6.19). The speed of rotation is

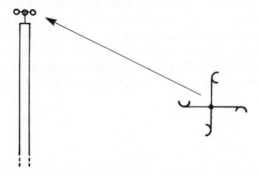

Fig. 6.19 The rotating cup sensor for wind velocity. The rotating arms can be mounted on the shaft of a tachogenerator or synchro for a direct electrical output.

proportional to wind speed, and is most conveniently measured by using a tachometer or a rotary digitiser connected to electronic circuits. The application of electronic methods to anemometers has increased the precision of these sensors in comparison to the older types in which the vanes were coupled to an indicator by a revolving cable.

Hot-wire methods are applied to gas movement as well as to gas pressure (see Chapter 1). A wire that is exposed to the moving gas is heated by the passage of a current. The temperature that the wire reaches depends on the cooling effect of the moving gas, so that the resistance of the wire can be used as a measure of its temperature and so of the gas velocity. The relationship between wire resistance and gas flow is not a simple one, and a sensor of this type needs to be calibrated against a more fundamental measurement. In addition, the calibration will need to be repeated if the composition of the gas is changed at any time.

The application of electronics to car engines has created a requirement for a number of gas sensors, particularly flow, temperature and composition sensors. Gas flow can very often be sensed by simple vane-deflection methods, because linearity is seldom an overriding requirement. Gas composition usually means exhaust gas oxygen content, which is a critical factor that measures the efficiency of combustion. The oxygen content of exhaust gas is sensed by using metal oxide 'cells' for which zirconium oxide or titanium oxide are the most suitable types. When either of these oxides is sandwiched between platinum electrodes, an EMF will be generated whose size depends on the availability of oxygen ions surrounding the sensor. By allowing the hot exhaust gas to pass over the cell, then, the amount of oxygen in the exhaust can be sensed, and so the efficiency of combustion assessed. The important feature of such a sensor for use in cars is that the change in combustion conditions from lean (too much oxygen) to rich (too much fuel) causes a very large change of oxygen content (a ratio of 10^{14}, for example) so that the output of the sensor can change very sharply in the range 20 mV to 1 V.

In general, sensors for gas composition rely either on cells whose output is an EMF, or on the effect of the absorption of the gas on a quartz crystal that is part of an oscillating circuit. These types of transducers are very specialised, and many of the cell types make use of the catalytic action of platinum or palladium in thin films. All such sensors are very susceptible to 'poisoning' by unwanted contaminants in the gases, and particularly metallic contaminants such as lead.

Viscosity

The viscosity of a liquid or gas is the quantity that corresponds to friction between solids, and which determines how much pressure will be needed to drive the liquid or gas through a pipe. In addition, some chemical reactions result in very marked changes in viscosity, so that sensing methods are needed for liquids that are static and not flowing. For almost all liquids, the viscosity, as measured by a coefficient of viscosity, decreases as the liquid is heated, but

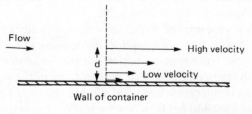

The velocity gradient is the quantity $\dfrac{\text{change of velocity}}{\text{distance between layers}}$

The viscosity of the liquid is defined as $\dfrac{\text{pressure}}{\text{velocity gradient}}$

Fig. 6.20 The definition of viscosity of a liquid in terms of the pressure and velocity gradient in a moving liquid. The velocity gradient cannot be measured directly.

for gases the reverse is true, and this is one reason why it is so much more difficult to deal with hot gas exhausts as compared to cold gas intakes.

Figure 6.20 shows the definition of the coefficient of viscosity in terms of the force that makes one layer of a liquid move faster than another. This force cannot be measured directly, but from the definition we can calculate the amount of force needed to spin a solid cylinder in a liquid, the amount of pressure needed to deliver a given volume per second of liquid through a pipe, or the rate at which a dense sphere will fall through the liquid. The traditional methods of measuring viscosity are based on such principles (Fig. 6.21), but

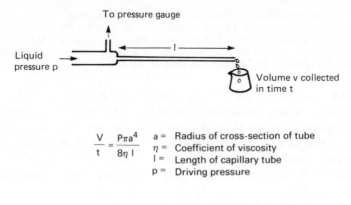

$$\frac{V}{t} = \frac{P\pi a^4}{8\eta\, l}$$

$a =$ Radius of cross-section of tube
$\eta =$ Coefficient of viscosity
$l =$ Length of capillary tube
$p =$ Driving pressure

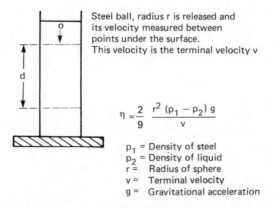

Steel ball, radius r is released and its velocity measured between points under the surface. This velocity is the terminal velocity v

$$\eta = \frac{2}{9}\frac{r^2\,(p_1 - p_2)\,g}{v}$$

$p_1 =$ Density of steel
$p_2 =$ Density of liquid
$r =$ Radius of sphere
$v =$ Terminal velocity
$g =$ Gravitational acceleration

Fig. 6.21 Traditional methods for measuring liquid viscosity values.

none of them is appropriate for electronic purposes apart from the pressure difference method which can be important in the control of continuous flow processes. Note that the methods of sensing liquid flow by means of the pressure difference across a restriction all assume that the viscosity of the liquid is constant – any change of viscosity will be read as a change of velocity.

Electronic sensors for viscosity can make use of the damping of mechanical

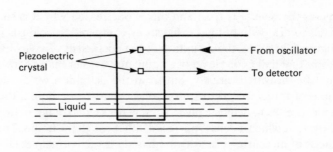

Fig. 6.22 The arrangement of piezoelectric crystals on a vibrating rod. One crystal is driven from an oscillator and the frequency is adjusted to make the vibration resonant, as shown by the peak output from the second transducer. This peak will be considerably damped if the rod is immersed in a viscous liquid, and the change in amplitude is a measure of the viscosity, if all other conditions are equal.

oscillations by a viscous liquid or gas. The principle is illustrated in Fig. 6.22 which shows a plunger that can be vibrated by a piezoelectric transducer, and whose amplitude of vibration is sensed by another transducer. If this arrangement is part of an oscillating circuit, then the damping effect of a liquid on the amplitude of oscillation will depend on the viscosity of the liquid, and can be sensed electrically as the amplitude of the signal from the receiving transducer.

Chapter Seven
Environmental Sensors

Environmental quantities

In the course of the preceding chapters, sensors for many quantities such as wind velocity which are of environmental importance have been mentioned. This chapter is concerned with some types of quantities and sensors which have not been covered previously and which are all in some way related to the environment. Some of these quantities need to be sensed in connection with industrial processes, and a few are of considerable importance in all aspects of life, but others are decidedly specialised.

All the devices that are considered in this chapter are sensors or measuring instruments, because none of the quantities is an energy form that permits conversion. There are, of course, natural forms of energy from which efficient transducers would be useful, but the low concentration of energy (the amount of energy per square metre or cubic metre of transducer) is very much against the use of such devices. It may seem attractive at first sight to generate electricity from wind or waves, but the size of any useful generator is daunting and it is most unlikely that our multitude of amateur environmentalists would ever permit such monstrous contraptions to be built. Practical generators must make use of processes such as combustion and nuclear fission which have a high energy concentration, and though the efficiency of conversion by way of steam may seem low at around 40%, it is very much better than is obtainable from other types of transducer systems, few of which ever look like bettering 10% with any degree of reliability.

Time

The measurement of time is fundamental to many industrial processes as well as to civilised life in general, but it is only comparatively recently that electronic timekeeping has become important. Our units of time are based on the solar year, the average time that the Earth takes for one revolution around the Sun, which is about 365.25 days. Astronomical determination of time is not exactly an everyday practical method, so that clocks and watches have been developed for convenience, and have traditionally relied on mechanical oscillations for a stable standard of frequency. Extraordinary efforts have been made in the past to achieve high precision of timekeeping because of its importance in navigation, but the limitations of a mechanical system have always determined what could be achieved. Really precise timekeeping has only now come about because of the application of electronic methods.

Electronic time measurement makes use of the crystal-controlled oscillator. The piezoelectric properties of the quartz crystal make it an excellent vibration transducer, and if the crystal is cut so that it resonates mechanically, it will act electrically as if it were an electrical resonant circuit with practically no losses. This implies that the frequency is very stable and that the oscillation can be maintained with very little energy input. For domestic electronic clocks and watches, this stability is sufficient, so that a cheap digital watch is a better timekeeper than the most expensive mechanical timepiece of a previous generation. For much more precise timing, the crystal should be kept at a constant temperature, and crystal standard-frequency generators have been built to this specification for some considerable time, since long before digital counting and displays were available. For the utmost stability, maser oscillators are now available.

The display of time or the measurement of time intervals makes use of digital methods, counting the waves from the crystal-controlled oscillators. If the system is to be used as a timer, a gate circuit must be added (Fig. 7.1) so

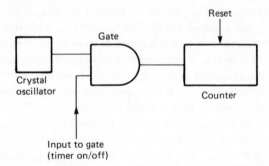

Fig. 7.1 The basis of an electronic timer system, using an oscillator, gate and counter/display.

that the counting can be switched on and off. The digital counter then starts when the gate is switched on and continues until the gate is switched off, with the count number being held (whether it is displayed or not) until another count is needed. If the gating is placed between the crystal oscillator and the counter, then the resolution can be to one oscillation, so that the frequency of the oscillator determines how precise the timing will be. A frequency of around 4 MHz is commonly used for crystal oscillators, allowing a resolution of around 250 ns. If the gating and counting circuits can use high-speed ECL ICs, then much higher master oscillator frequencies can be used.

Moisture

The presence of moisture in gases or solids often needs to be sensed or measured. A high moisture content in a gas (high humidity) will cause condensation when the gas is cooled, and this can have the effect of depositing liquid in pipes, causing blockages. Conversely, a gas whose humidity is very low can absorb moisture from joints in a pipe, causing the joints to dry up and leak. The moisture content in solids may be desirable, as in the preservation of antique furniture or the working of textiles, or undesirable when mould starts to appear on brickwork. All of this assumes that the moisture is water, but in some specialised applications the presence of other liquids may need to be sensed, and such applications require much more specialised equipment.

The presence of moisture in a gas is termed humidity, and the absolute humidity of a gas (usually air) is the mass of water per unit mass of gas. This absolute humidity is the quantity that needs to be known if the amount of water that can be condensed from a gas has to be determined. The amount of water that can be contained in a gas is limited, and the maximum humidity attainable is called the saturation humidity. This figure depends heavily on the temperature, being very low at low gas temperatures (around 0°C) and very high at temperatures approaching the boiling point of water (100°C).

For many purposes, the relative humidity is a more important value than the absolute humidity. The relative humidity at any temperature is the absolute humidity divided by the value of saturated humidity at that temperature, usually expressed as a percentage (Fig. 7.2). A relative humidity value of 50% at 20°C, for example, means that the air contains half of the quantity of water that would be needed to saturate it at this temperature. Calculations of absolute humidity from values of relative humidity are never simple because tables of saturated humidity always quote vapour pressures, and Fig. 7.3 shows how these values can be used to obtain absolute figures.

Another measure of relative humidity is the 'dew-point'. When a surface is cooled in contact with a gas, a temperature will eventually be reached when the gas deposits water onto the surface by condensation. This surface

---

Relative humidity % $= \dfrac{\text{actual humidity}}{\text{saturated humidity}} \times 100$

Example: One cubic metre of air is dried out and found to contain 0.006 kg of water. The amount of water in saturated air would have been 0.017 kg.

Relative humidity is $\dfrac{0.006}{0.017} \times 100 = 35.3\%$

Fig. 7.2 The relationship between relative humidity and absolute humidity. The absolute humidity is normally measured as kilograms of water per cubic metre of air.

Humidity $= \dfrac{0.00217\,\rho}{T}$

Humidity figure in kg of water per m³ of air
  ρ = pressure of water vapour (Pascals)
  T = Kelvin temperature (°C + 273)

Example: Pressure of water vapour in saturated air at 30°C is 4240 Pa

Humidity is $\dfrac{0.00217 \times 4240}{303}$ = 0.030 kg
                                    (30 g)

Fig. 7.3 Relating the water content in kilograms per cubic metre to the quoted figures of pressure of water vapour.

temperature is called the dew-point temperature, and its significance is that it corresponds to the temperature at which the gas would be saturated. If the dew-point temperature is known, then, the vapour-pressure of the water in the gas can be found from tables, and either the relative or the absolute humidity calculated, as illustrated in Fig. 7.4.

Electronic methods of measuring humidity are based either on dew-point or on the behaviour of moist materials. Of these, methods that utilise moist materials are by far the simplest to use, though they need to be calibrated at intervals if they are used for measurement as opposed to sensing purposes. The simplest method is a very old one, using the principle of the hair hygrometer. A human hair, washed in ether to remove all traces of oil or grease, is very strongly affected by humidity, and its length will shrink in dry conditions and expand in moist conditions.

A very satisfactory relative humidity sensor can therefore be constructed by using a slightly tensioned hair along with any of the standard methods for detecting length changes, such as the linear potentiometer, LVDT, capacitive (with oscillator) and so on. As usual, the use of a capacitive gauge allows the reading to be made in terms of change of frequency and makes it easy to treat the information by digital methods. The amount of force on the hair must be very small, so that whatever method of sensing length changes is used must not place an excessive load on the hair. The system can be calibrated either

Temperature of air   20°C
Dew-point   9°C

∴. Air is saturated at 9°C. The pressure of water vapour is therefore 1150 Pa (from tables of saturated vapour pressures).
  The saturated vapour pressure at 20°C is 2340 Pa (from tables).
∴. Relative humidity $= \dfrac{1150}{2340} \times 100$ = 49%

Fig. 7.4 How the relative humidity is found from dew-point readings. The method depends on having tables of saturated vapour pressure. These are easily obtainable for water and air, but are not so easy to obtain for other liquids and gases.

against another indicating hygrometer, or against a standard chemical humidity measurement (Fig. 7.5).

Though the hair type of hygrometer is capable of surprisingly good results, the lithium chloride type is more directly suitable for electronic purposes. Lithium chloride has a very high resistance in its dry state, but the resistance drops considerably in the presence of water, and over a reasonable range the resistance reading can be used to measure relative humidity. The lithium chloride cell is made part of a measuring bridge, and the output is used to drive a meter movement, trigger a switch action at some critical value, or is converted to digital form for display.

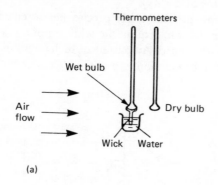

(a)

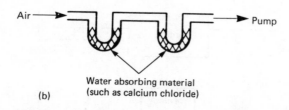

(b)

Fig. 7.5 Standard methods for humidity. The wet-and-dry thermometer method (a) is not very precise, and depends on the use of tables. Its merit is simplicity. The chemical absorption method (b) draws a metered amount of air past a desiccating material which is weighed before and after the process. The weight difference is the weight of water present in the air. This latter method is precise but slow and difficult to set up.

Other effects can be used to determine the moisture content in absolute terms. The presence of moisture in air alters its value of permittivity, so that the capacitance between two fixed metal plates in air will change very slightly as the moisture content of the air changes. This as usual can be sensed in the form of a change of frequency of an oscillator, but the readings have to be corrected for changes of temperature, since the dimensions of the plates will change as the temperature around them changes. Another effect that can be utilised is the heat conductivity of the air which will alter as the moisture content is altered. The most useful effect, particularly for remote measurement, is microwave measurement.

Microwaves at a frequency of 2.45 GHz are absorbed strongly by water (the principle of microwave cooking), so that if microwave signals at this frequency are beamed across air and returned from a fixed reflector, the amount of returned signal will depend very considerably on the water content in the intervening air. This method has the considerable advantage that it senses the average water content along a path in air, which can be quite large. The system needs to be calibrated, but once calibrated, the amplitude of the returning signal can be used as a measure of humidity (the greater the returned signal, the less the humidity).

Dew-point methods can provide fairly precise measurements of either absolute or relative humidity, but require the addition of a microprocessor system if direct readings are needed. An outline of such a system is illustrated in Fig. 7.6. This uses a Peltier junction as a cooling element, using the well-

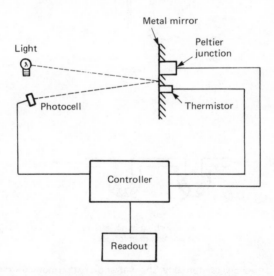

Fig. 7.6 An outline of an electronic relative humidity detector system using a Peltier junction for cooling, a thermistor to detect temperature and a photocell to detect the formation of dew. It is easier to use two thermistors in a wet and dry system, but the results are much less reliable.

known principle that a pair of junctions between metals that provides an EMF for an applied temperature difference will also operate in reverse, providing a temperature difference when connected to a supply. When semiconductor junctions are used, the change in temperature when current is passed can be large enough to allow one junction to attain temperatures lower than 0°C while the other junction is held at room temperature. Since the effect is current-controlled, the rate of fall of temperature can be controlled electronically.

The Peltier junction is therefore used to lower the temperature, and a thermistor measures this temperature. A beam of light is reflected from the mirror-surface of metal attached to the junction, and a photocell senses the reflected light. At the dew-point, the mist on the surface causes the amount of reflected light to drop sharply, and the amplifier connected to the photocell operates the reading gate of the system, so that the output from the thermistor temperature gauge is converted to digital form and stored. The microprocessor system also stores the output from another thermistor gauge which senses room temperature, and its ROM contains a set of table values for the temperature readings and for relative or absolute humidity readings (the absolute humidity can be found only if the volume of the space is known). The method of obtaining relative humidity has been outlined in Fig. 7.4.

Moisture in solids

Moisture in solids can be sensed in terms either of the conductivity of the materials, changes in permittivity, or of the absorption of microwaves. For sensing the presence of moisture in materials of comparatively fixed composition (and even where some variation occurs, as in masonry), a simple resistance reading between connectors set at a fixed distance is often all that is required. For measurement purposes, readings have to be calibrated, and calibration is a long and tedious procedure which requires samples with various moisture levels to be checked for resistance, then weighed, baked to remove moisture and weighed again. The difference in weights shows the moisture content which can be expressed as a percentage of original weight of material.

A very common requirement is for moisture content of soil. This is very seldom required to be precise, which is just as well, because the relationship between the resistance of the soil and its moisture content is not a simple one. A simple resistance indication, however, is enough to tell whether a plant needs watering or not, which is the main reason for using moisture indicators. For civil engineering purposes, the use of a soil-resistance moisture meter is only a preliminary indicator, and a complete soil sample analysis would be needed before any decisions on the suitability of soil for foundations were made.

Acidity/alkalinity

The acidity or alkalinity of water is an important factor for water suppliers, and also for all users of water such as chemical plants, generating stations, agriculture and horticulture. The acidity or alkalinity of water is measured on the pH scale, on which perfectly neutral water has a pH value of 7, fairly strong acid solutions have a pH of 2 and fairly strong alkali solutions have a pH of 12. The basis of this scale is the relative amount of free hydrogen ions in the water, and is outlined in Fig. 7.7.

The sensing of pH can be fairly simple if all that is needed is an indication of a change in the ionisation. Perfectly neutral water, with a pH of 7.00, has a very high resistivity value, but any trace of ionisation will cause the resistivity to drop very sharply. Some care is needed in the measurement of resistivity, however, because most metals dissolve to some extent in water, creating ions and causing a resistivity change. Metals such as platinum or palladium are best suited to this type of use, but this type of indication does not show whether the conductivity is due to hydrogen ions (acidity), hydroxyl ions (alkalinity) or metallic ions (contamination).

Pure water: H−O−H hydrogen atoms attached to each oxygen atom. One in 10^7 is dissociated to H^+OH^- to form ions. This corresponds to pH = 7

Acid: H^+Cl^- almost completely dissociated (separated) into ions. If the presence of the acid means that H^+ is present to the extent of 1 in 10^3, this corresponds to pH = 3

Alkali: Na^+OH^- almost completely dissociated into ions. The OH^- ions suppress any dissociation of water molecules, so that 1 in 10^{12} might be dissociated. This corresponds to pH = 12

Strictly   pH = − $\log_{10}$ (hydrogen ion concentration)

Fig. 7.7 The basis of the pH scale for acidity and alkalinity of water solutions.

The standard electrical method for reading pH depends on the glass-electrode system, illustrated in Fig. 7.8. The glass bulb is very thin and contains a mildly acidic solution which is a good conductor. A platinum

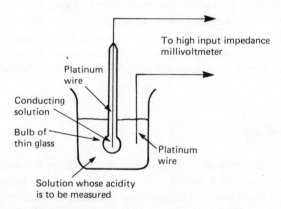

Fig. 7.8 The glass electrode method of measuring pH. The output is a voltage which is proportional to pH, but the very high impedance of the glass requires the use of an FET (or valve) input stage for the millivoltmeter.

contact is inside the bulb, and another platinum contact is immersed in the water whose pH is being measured. The EMF that exists between the glass electrode and the external platinum wire is then a measure of pH value in the water. The resistance of this cell is very high, of the order of 100 MΩ, so that high input-impedance MOS DC amplifiers are needed to read the few millivolts of output.

The glass-electrode pH measuring system is fragile, and the readings are easily upset if the glass is allowed to dry out or become stained, but of the available methods it is the most reliable for any kind of use outside laboratory conditions. Calibration is easily carried out, because solutions of standard and constant pH value (buffered solutions) are easily obtained.

Wind-chill

The effect of wind on heat-loss has been known for a long time, but wind-chill factors have only recently been quoted in weather forecasts to the general public. The principle is that an object located in still air will lose heat comparatively slowly, because the air itself acts as a thermal insulator. Even for a comparatively large temperature difference, then, the rate of loss of heat can be low. When the air is moving, however, its cooling effect is much greater. The layer of air that is in contact with the warm surface is constantly being removed and renewed, taking its heat with it, so that the effect of moving the air is the same as the effect of being immersed in much colder, still air. The wind-chill temperature expresses the temperature of still air that will provide the same rate of cooling as the moving air at a higher temperature. The factor is often misunderstood – if the air temperature is 8°C and the wind-chill temperature is 2°C, then the temperature of an object in the air will drop to 8°C, but the rate at which the temperature drops will be faster, as if the heat were being lost to a 2°C air temperature.

Wind-chill is easily amenable to electronic measurement, and Fig. 7.9 shows one system. The thermistor measures the temperature of the sensor, which is maintained at a constant temperature by a heating element whose current is measured. In still air, the amount of current required to keep the

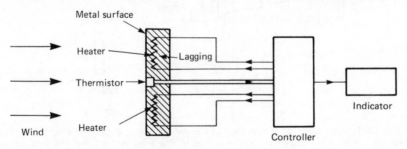

Fig. 7.9 The wind-chill indicator system.

sensor at a constant temperature will be low, because the loss is comparatively low. In moving air, considerably more current is required to maintain the temperature constant, and this change of current is a measure of the wind-chill factor. A more practical arrangement uses one sensor in the moving air and one kept in still air at the same air temperature, comparing the current readings so as to indicate the wind-chill. Calibration is needed, and this can be done by taking readings for several values of temperature difference between the sensors.

Radioactive count rate

An environmental factor that has sprung into prominence in the last few decades is the radioactive count level, and this figure has been used in such a way as to prove that a little learning can be a remarkably dangerous thing. The quantity that is measured in these readings is the number of ions produced per second by all types of radiation and by particles from radioactive materials. Since the whole Earth consists of materials that are to some extent radioactive, and is constantly bombarded with radiation from the Sun and other stars, there is no place that is free of radiation. In addition, the normal level of radioactivity varies very considerably from one place to another, and is particularly high where there are old granite rocks, or in the presence of deep-mined materials. One very revealing test is to measure the radiation count downwind of a coal-fired power station or on moors of granite rock. Most detectors can be driven completely off-scale by the radiation from any pre-war luminous watch, because these used radium for the luminescence.

The basis of many counters of ionising radiation or particles is the Geiger-Muller tube. This (Fig. 7.10) consists of a tube that contains a mixture of gases

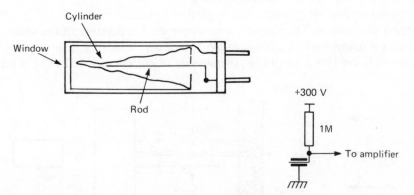

Fig. 7.10 The construction of one form of Geiger-Muller tube, and the electrical circuit. The presence of an ionising particle or ray in the tube ionises the gas and causes conduction until the ions are absorbed. The output is one brief pulse for each ionising 'event'.

(usually the inert gases of the air, such as krypton) at a low pressure, about 1/80 of atmospheric pressure. The electrodes in the tube are maintained with a voltage difference of about 400 V, and in the presence of radiation or ionising particles, the gas in the tube is itself ionised, allowing a brief pulse of current to pass. The current is brief because the gas mixture contains traces of bromine or iodine which have the effect of neutralising the ions rapidly so that the gas does not continue conducting after the cause of the ionisation has passed.

The output from the tube is taken across a load resistor and consists of a pulse for each particle when the particles do not arrive in great profusion. The pulses can be amplified and used to operate a counter, a rate-meter, or a loudspeaker to give the clicking noise beloved of films on radioactivity. The count rate is obtained by passing the pulses into an integrating circuit whose DC voltage is then proportional to the rate at which the pulses arrive. The GM counter, however, detects only the ionisation caused by radiation or particles – it does not indicate the cause of the ionisation, nor is it equally sensitive to all causes of ionisation.

Natural sources of radioactivity can produce ionisation from three causes. The first cause is the alpha-particle, which is an ionised nucleus of the gas helium. This has a comparatively large mass and very strong ionising effect (and a correspondingly large effect on living cells), but is absorbed strongly in all materials, including air, so that its range from its source is usually only a fraction of a millimetre. Only GM tubes with a very thin entry window (usually mica) can detect alpha-particles, and only when the window end of the tube is held against the source of the particles.

The second type of particle is the beta-particle, which is the familiar electron. This particle ionises materials quite efficiently but is very much smaller than the alpha-particle so that its effects on living cells are much less. The range in air can be several centimetres, but a sheet of paper is enough to stop electrons from all but a very active source. Electrons are very efficiently detected by the GM counter.

The third type of radiation is the short-wavelength ray, such as the gamma rays from radioactive materials and the cosmic rays from outer space. These are only weakly ionising, but are immensely penetrating and require several feet of concrete or several inches of lead for screening. Because these rays have a rather weak ionising effect, they are not efficiently detected by the GM counter. These rays, however, are the most dangerous to life.

As well as these natural sources of radiation from the Earth itself and from outer space, the use of natural radioactive materials in concentrated form (as in nuclear power sources) gives rise to other particles. The main particle of interest is the neutron, which has virtually no ionising effect, since it is electrically uncharged, and which therefore does not affect the GM counter unless the bombardment is dense. The alternative is to use a GM tube which contains a vapour that will absorb neutrons and give out electrons. Neutrons are very penetrating and harmful, but difficult to detect. Shielding can make use of water, paraffin wax or other low-density materials.

The main alternative to the GM type of counter is the scintillation counter. The basis of this is a crystal which will give off a faint flash of light when affected by an ionising radiation. The crystal is kept in a dark space and the faint light is detected by a photomultiplier (see Chapter 3) so that the pulses from the photomultiplier can be counted. Several types of sensitive crystals can be used, so that one type of radiation can be detected rather than another. This makes it possible to estimate the relative contribution that different types of radiation make to the whole.

One of the main problems of using detectors like the GM counter is to determine what amount of the reading is due to the presence of an unwanted radioactive material. The continual bombardment of ionising particles and radiation from the Earth and from space produces a background count, and this count has to be determined, often over a long period, and subtracted from the count obtained in the presence of suspected radioactive material. Low-level radioactive material, such as laboratory glassware and clothing used in the radiological industries, have a count level so close to the natural background level that it is difficult to establish the level of contamination. Given the choice of having a nuclear dump or a heap of manure next to your house, it's better to plump for the nuclear waste!

Surveying and security

The measurement of distances that may not allow for direct methods is a problem that is often encountered in surveying, and even the measurement of room sizes in an occupied house can be difficult. This type of problem can be tackled by ultrasonic sonar methods, and similar methods can also be used in security systems to warn of movement inside the space covered by the sonar beam.

The sonar distance gauge principle is illustrated in Fig. 7.11. A beam of ultrasound, usually in the 40 KHz frequency range, is sent out from a transducer, usually as short pulses. At this frequency, the wavelength in air of the ultrasound is around 7.5 mm, so that the resolution of the system is of this order. The reflected beam is picked up, in this example by another transducer, and the time interval between sending and receiving is converted into a reading of distance, using either analogue or digital methods. The same transducer can be used both for transmitting and for receiving if the time interval between pulses is large enough to allow for the maximum delay of the beam in the largest space that will have to be measured.

The system requires some care if it is to be used correctly, because the beam is invisible, making it impossible to be sure where the main reflecting point is located. False readings can be obtained if the beam is angled in a room, so that the distance that is measured includes more than one reflection (Fig. 7.12).

For security systems, the ultrasonic beam is usually continuous, with a

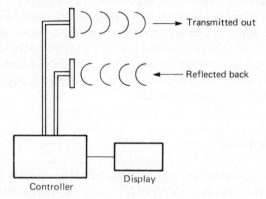

→ Transmitted out

← Reflected back

Controller

Display

Fig. 7.11 The sonar distance gauge. Pulses are generated in the controller and used to send waves from the transmitter transducer. The received echo signals are amplified, and the time delay between these and the transmitted pulses used to compute distance. When two transducers are used, as illustrated, the signals can be continuous rather than pulsed, and phase change measured rather than time. Pulsed systems can use a single transducer, switched between transmitting and receiving.

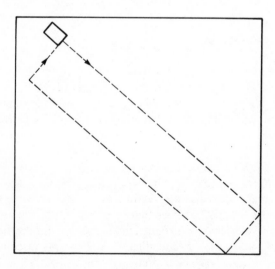

Fig. 7.12 The receiving transducer is not particularly directional, so that a reflected path such as is illustrated here can cause false readings.

separate receiver transducer. Inside a room or other closed space, a complex pattern of standing waves is set up and the signal from the receiver transducer is compared with a phase-shifted signal from the transmitter, and the amount of phase shift and amplitude adjusted until the signals are of equal amplitude and phase. If these signals are then applied to a phase-sensitive comparator, then any change in the phase of the received signal will cause an output from the comparator which can be used to trigger an alarm. Such a phase change can be caused by an intruder, but the sensitivity of the system is often such that the alarm can be triggered by cats, mice or pieces of paper moving in a breeze. The more elaborate systems therefore use amplitude discrimination as well as phase shift to trigger the alarm.

Animal fat thickness

One of the more specialised uses of sonar distance meters has been in measuring the thickness of the layer of fat under the skin of animals, pigs in particular. The speed of sound in a fatty layer is different from the speed in the more dense meat, so that an ultrasonic beam will be reflected from the place where fat ends and meat starts. By making the usual measurement of time elapsed between sending the ultrasonic pulse and receiving the reflection, the thickness of the fat layer can be measured painlessly.

Chapter Eight
Scientific and Engineering Requirements

Unusual measurements

Laboratory work, whether comparatively routine in nature or involving new ideas, often requires sensing and measuring systems that have not been covered by any of the examples up to this point. In many cases, this does not imply that completely new sensors have to be developed, because an unusual application can often be catered for by using a combination of existing sensors. Research laboratories, however, generally have to construct their own sensing and measuring systems because the nature of the work implies that nothing is available. In this chapter, we shall consider some types of sensors and measuring methods that are used in laboratory work, in engineering development and in some types of geological work, but which find little or no application outside these uses.

Permittivity

The permittivity of an insulating material determines its effect on the capacitance of a pair of conducting plates that sandwich the material. The usual measurement of permittivity is as relative permittivity, meaning the permittivity of the materials divided by the permittivity of space (a vacuum). This quantity was formerly known as dielectric constant. The permittivity of a vacuum is close enough to the permittivity of air to allow this amount to be used except for the most precise measurements. The measurement can be for the purpose of determining how effective a new material (usually a plastic or ceramic material) would be when used as a dielectric for capacitors, but it can also be used in assisting to determine the structure of a material.

The outline of the system for measuring relative permittivity is shown in Fig. 8.1. The parallel-plate capacitor is constructed with circular plates and a 'guard-ring' to offset the effects of stray capacitance at the edge. The value of capacitance is measured, using a capacitance meter, with and then without the dielectric, maintaining the same spacing between the plates. The ratio of capacitance values is taken as the relative permittivity of the material.

The principle of the capacitance meter is shown in Fig. 8.2. The capacitor

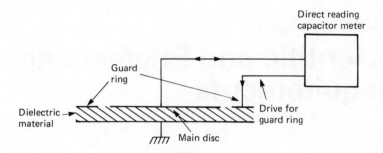

Fig. 8.1 Measurement of relative permittivity. The system uses a circular disc capacitor, with a guard ring whose action is to suppress stray capacitance. The voltage on the guard ring is identical to the voltage on the main plate, but no measurements of capacitance are taken from the guard ring.

is alternately charged from a constant-voltage source and discharged through a meter or into an integrating counter. The switching is carried out by a MOS array, and for high sensitivity has to be done at a high frequency, several MHz. The capacitance is obtained as shown in the example from a reading of current, and the whole system can be made direct-reading.

In addition to the relative permittivity of a capacitor dielectric, its loss factor may have to be assessed. This is a very much more difficult measurement unless the loss factor is large, though for some purposes (such as soil sample analysis) large loss factors are normal. The measuring method that is used is a bridge in which the capacitor with the dielectric under test is

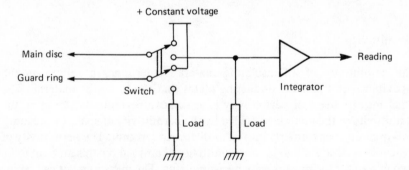

Example: Switch operates with 50% duty cycle at 1 MHz. For the capacitance this means that the charging and discharging is each completed in 0.5 µs. If the voltage is 10 V and the capacitance is 50 pF then the charge is 500 pC, which is discharged 10^6 times per second. This corresponds to a current average of $10^6 \times 500 \times 10^{-12} A = 5 \times 10^{-4} A$ or 0.5 mA, an easily measurable quantity.

Fig. 8.2 The principle of the direct reading capacitance meter and how the output current is related to the capacitance value. The switching device would normally be a MOSFET bridge, though reed relays can be used for large capacitor discs.

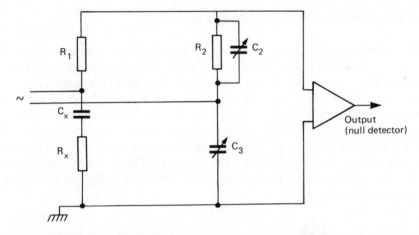

Fig. 8.3 The Schering bridge method for measuring capacitance and leakage resistance. Modern methods for leakage resistance charge the capacitor and measure the time constant from the rate of fall of voltage.

balanced against a standard low-loss capacitor and a phase-shift resistor, using the circuit of Fig. 8.3, the Schering bridge, or a more specialised type.

Electrostatic charge sometimes has to be measured by electronic methods, and the simplest method is to measure the potential at various distances from the charge – the relationship between potential and charge is shown in Fig. 8.4. The 'connection' between the voltmeter and the air is established by using a flame to ionise the air around the detector. The voltmeter must have a very high input impedance, and a MOS type is normally used to replace the older valve voltmeters that were once common in measuring laboratories. Two determinations of potential at different distances are usually needed to establish the amount of charge, though this method will not be suitable if the charge leaks away rapidly. Note that there is no electrical connection between the charged object and the voltmeter: the potential is due to the field set up by the charge.

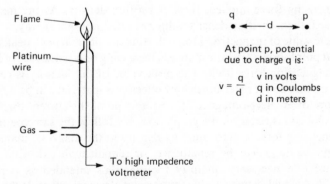

At point p, potential due to charge q is:

$$v = \frac{q}{d}$$

v in volts
q in Coulombs
d in meters

Fig. 8.4 The flame electrode method of measuring the potential at any point in the air. Obviously this is unsuitable if the air contains flammable vapours. The relationship between charge, distance and potential is also shown.

Permeability and magnetic measurements

Permeability is analogous to permittivity, and its effect on the inductance of a coil is illustrated in Fig. 8.5. The important difference, however, between these types of quantities is that permeability and relative permeability are not constants for a given material. We can quote a single figure of relative permittivity for a sample of mica, but for a sample of steel the value of relative permeability is not a constant, though the maximum possible value may be fairly constant. Relative permeability values depend on the previous history of the material, in the sense that if the material has been in a magnetic field, then its value of relative permeability will be affected by the fact that it has been magnetised.

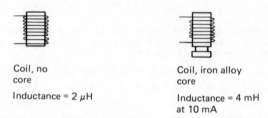

Coil, no
core

Inductance = 2 μH

Coil, iron alloy
core

Inductance = 4 mH
at 10 mA

Fig. 8.5 The effect of relative permeability on the inductance of two otherwise identical coils. Whereas values of relative permittivity of 2 or above are comparatively rare, most ferromagnetic materials have relative permeability values which can be measured in thousands or tens of thousands.

Rather than quote a single figure to describe the behaviour of a material in a magnetic field, then, we have to quote a number of figures, and these are illustrated in Fig. 8.6 with reference to the magnetic hysteresis curve of a material. The graph shows the magnetic flux density (B) in the material plotted against the ampere-turns of magnetising force in a coil surrounding the material. We assume that the material starts unmagnetised, so that with no magnetising force applied, there is no flux density. As the material is magnetised, then, the flux density value rises in a non-linear way.

At some value of magnetizing force the flux density reaches a peak and will from that point show very little increase, increasing only at the rate of increase of the magnetising force itself. The peak value of flux density is called the saturation flux, and it determines how effective a material can be when used as the core of an electromagnet. If from this point the current through the magnetising coil is reduced, the graph does not follow the same path. When the magnetising force is zero, the flux density in the magnetic material will have some value called the remanence, or remanent flux density. A high remanence is a necessary quantity for materials intended as permanent magnets, but is highly undesirable for cores of electromagnets. If the current through the magnetising coil is now reversed and increased, a point will be reached when the flux density in the sample of material is zero. The amount

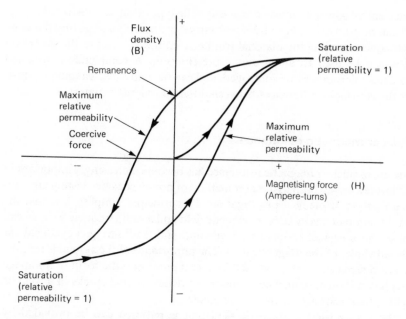

Fig. 8.6 The B-H hysteresis curve for a magnetic material. The value of relative permeability is obtained from the slope of the curve, and is unity at saturation. The graph also shows the value of remanence and coercive force.

of magnetising force needed to achieve this is called the coercive force, and a high value of coercive force means that a material will be difficult to demagnetise, a desirable feature of, for example, material used for recording tape.

The use of Hall-effect sensors allows flux density to be measured even for small samples of material, so that evaluation of these magnetic quantities is

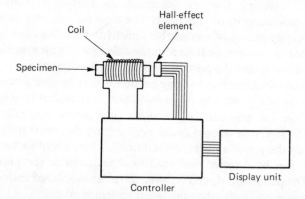

Fig. 8.7 Outline of an automatic hysteresis curve tracer. A timebase generates the rising and falling currents through the coil, and this same waveform is applied to the X-plates of the cathode ray tube display. The output of the Hall-effect sensor is amplified and used to provide the Y-plate signal.

much easier now than it was previously. The type of set-up illustrated in Fig. 8.7 can be used to display a hysteresis curve on an oscilloscope, and the same technique of cycling the material from one saturation level to the other and back can be used to allow digital presentation. A completely automated system is microprocessor controlled, and can be made auto-ranging to allow for the enormous differences between different magnetic materials.

Nuclear magnetic resonance

The use of nuclear magnetic resonance has become increasingly important in the last few years, both as a major method of non-destructive testing and also as a method of medical investigation. The principle, simplified, is that the combination of a very strong magnetic field and a radio frequency signal can cause the nuclei of atoms to absorb and emit radio-frequency signals at certain values of the magnetic field. The importance of the system is that the received signals can be processed by a computer into the form of an image which can show a remarkable amount of detail, so that cracks in a weld or defects in a heart are visible on the image.

The very powerful magnetic field that is required can be provided by conventional electromagnets when the specimens are small, but for purposes such as head and body scanning, a superconducting magnet is required with a maximum field strength of about 0.5 T. For medical purposes, the RF power is about 100 to 200 W, and the frequency range of signals is 1 to 100 MHz.

Gravitational sensing

For mineral surveying, the tiny changes in the Earth's gravitational field caused by dense deposits of minerals can be detected and used to locate the deposits. Gravity meters of remarkable sensitivity have been available for a long time, but in recent years the addition of electronic sensors has improved both the sensitivity and the portability of gravitational meters.

The basic methods are the swinging pendulum and the spring balance. The time per swing of a pendulum depends on the gravitational acceleration, as shown in Fig. 8.8, and can be measured very precisely by allowing the pendulum to interrupt a light beam and sensing the interruption with a photocell. By the use of a quartz crystal timer, a very small variation in the swing time can be detected and read as a variation in the gravitational constant. The pendulum should be mounted in an evacuated enclosure, and should be swinging freely when the measurement is taken.

The alternative is to measure the changes in the deflection of a spring loaded by a mass. This calls for the sensing of very small changes in length, and a laser interferometer is the best method available. Once again, the whole

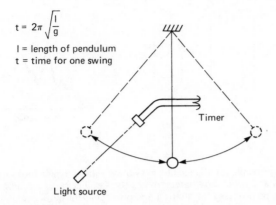

$$t = 2\pi \sqrt{\frac{l}{g}}$$

l = length of pendulum
t = time for one swing

Timer

Light source

Fig. 8.8 Measuring the time of a swinging pendulum with a quartz crystal clock. If the pendulum is in an evacuated enclosure, the time for one swing depends almost only on the value of the gravitational acceleration, g.

balance should be mounted in a vacuum to avoid the effect of air damping and air currents.

Spectrum analysers

Spectrum analysis means obtaining information on the amplitude of waves for each frequency of waves in a set, the spectrum of the title. Originally, the phrase applied only to the visible spectrum, but developments in wave detection led to infrared and ultraviolet spectrum analysers being developed, and the title is now also applied to instruments that analyse the radio frequency spectrum in a similar way. The principle of all types of spectrum analysers (Fig. 8.9) is that each frequency in the spectrum is selected by a filter and its amplitude measured. The display of the information can be on a cathode ray tube, or as a graph or a printout from an attached computer.

For the radio-frequency spectrum, the filtering action is carried out by tuned circuits, using LC circuits for the lower frequencies and cavities for the microwave region. The range of frequencies requires any radio-frequency spectrum analyser to operate in switched bands, because no single tuning circuit can operate over more than a small part of the useable range. For the visible spectrum of light, the filtering action is obtained by using a glass prism, using the principle that the angle through which a light beam is bent will depend on the frequency of the light.

Infrared and ultraviolet spectrum analysers cannot make use of glass prisms apart from the small part of the spectrum that lies next to the visible range. Materials such as quartz for ultraviolet and rock-salt for infrared can be used, and in the region where the longer-wave infrared meets the short-wave microwave range, prisms of paraffin wax are effective. The shorter the wavelength of the radiation, the more difficult it is to find a material for a

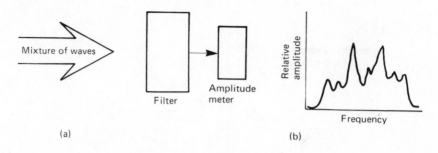

(a)

(b)

Fig. 8.9 The operating principle (a) of a spectrum analyser, and a typical output (b). For radio waves, the filters and amplitude detectors are the familiar radio circuits, but for optical frequencies prisms are used for filtering and photocells for detection.

prism, because the action of a prism depends on dispersion, meaning that the speed of the waves in a material depends on their frequency. The shorter the wavelength, the less is the interaction with the atoms of materials and the lower the dispersion. As we approach the X-ray frequencies, nothing is suitable as a prism and practically all materials are transparent.

Spectrum analysers in all classes are usually operated by electronic methods, and make use of electronic detection. For the radio frequency ranges, the filters can be electrically controllable and the filtered frequencies detected by a diode, so that the plotting of received amplitude for each frequency can be controlled by a microprocessor. The higher frequencies, of which the visible range is at the higher end, are dealt with using a servo-motor to rotate the prism and a detector such as a photocell to measure the amplitude of the received signal.

Index